MY FIRST SUMMER
IN THE SIERRA

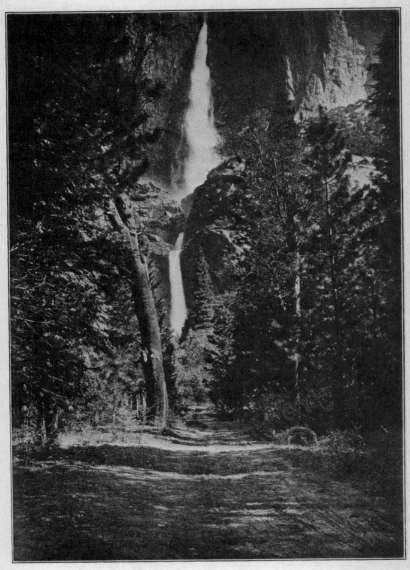

THE YOSEMITE FALLS, YOSEMITE NATIONAL PARK

MY FIRST SUMMER
IN THE SIERRA

JOHN MUIR

INTRODUCTION BY
GALEN ROWELL

A MARINER BOOK
HOUGHTON MIFFLIN COMPANY
BOSTON   NEW YORK

TO THE SIERRA CLUB OF CALIFORNIA
Faithful defender of the people's playgrounds

First published by Houghton Mifflin Company in 1911.
First Mariner books edition 1998.
Introduction copyright © 1998 by Galen Rowell

Library of Congress Cataloging-in-Publication Data is available.
ISBN 0-395-35351-3

Printed in the United States of America
QUM 10 9 8 7 6 5 4 3 2 1

CONTENTS

ILLUSTRATIONS

ILLUSTRATIONS

All other illustrations are from sketches made by the
author in 1869.

INTRODUCTION

MY first summer in the Sierra and John Muir's were not that different. A century apart, the "gentle wilderness" around timberline caught both our fancies. We reveled in the wildness of flowered meadows and indigo lakes, camped beneath quilted mountains of snow, ice, and granite slabs, thought some of the same thoughts and had some of the same feelings. Readers who have experienced the Sierra high country themselves will share my uncanny feeling that I have walked these pages with Muir, rejoicing at pristine streams cascading down from the snowy peaks, learning the names of the flowers and trees, and feeling the same emotions about wildness that he expressed so well so long ago.

While I'm aware of the pitfalls of relating my own experiences in the introduction to another man's writings, they are relevant here to illustrate the lasting meaning of Muir's life. Other nineteenth-century writers have left us only words from worlds long gone. Mark Twain's Old West and Rudyard Kipling's colonial India have undergone profound changes over the last century, yet John Muir's wild Sierra was still there at

INTRODUCTION

my feet in 1951 as well as on a crisp fall morning in 1997, when I repeated his solo climb of Cathedral Peak without encountering another human being.

In his later life Muir used the power of his fame and the muscle of his prose to help protect wildlands for posterity. He was a prime mover in the creation of the national park that now protects not only the immutable rock features of Yosemite Valley but also the fragile beauty of the high country he so vividly describes in these pages. He also founded the Sierra Club, to which he dedicates this book. Thus it should come as no surprise, as a second century turns over beyond the one of that memorable summer, that the book remains solidly in print and widely read.

Had Muir not made enormous efforts to preserve the wildlands surrounding his youthful adventures, this book might never have been published at all. A mystic might say it all came together because of ancient twisted karma, yet Muir's amazing destiny can be directly traced to right intentions practiced during his lifetime. The threads of experience that led him in 1911 to take out his well-worn notebooks from the summer of 1869 and revise them only slightly for publication can easily be traced in his life and letters.

In 1904, Muir was planning to take an extended

INTRODUCTION

trip through Asia and Europe, when he received
word that President Theodore Roosevelt wanted
to meet with him. The impulsive thirty-one-year-
old who wrote the notes for this book in 1869
would never have been invited to meet the president. The legendary sixty-five-year-old conservationist who did receive that honor was past the
peak of his amazing physical prowess, yet he was
still more fit than Roosevelt, who had ridden into
office on a reputation of great physical tough-
ness. Muir's first inclination was to say no to the
president, as he had to other famous people in
his younger years in order to follow his own star
into the mountains. In this case, however, Roosevelt suggested that the two of them sneak off
on foot and rough it for a few days somewhere in
Yosemite.

Three things helped Muir make up his mind.
First, Yosemite Valley was not then a national
park, although parkland surrounded it on all sides.
The valley had been held out to be managed by
the State of California, with far less protection
and more development than its natural splendor
warranted. Second, forests were being cut all
over California, and Muir saw a chance "to do
some forest good." But the final straw that con-
vinced him to postpone his overseas journey was
a personal message from the president: "I do not
want anyone with me but you, and I want to drop

politics absolutely for four days and be out in the open with you."

Muir agreed to come, but he did not drop politics. After the Rough Rider president and the old mountaineer discovered around the campfire that they were kindred spirits, Muir began preaching forest conservation. Roosevelt emerged from the natural cathedral of tall trees with a directive to the Department of the Interior to protect the forests all the way from Yosemite to Mount Shasta, hundreds of miles to the north. He also told Muir that he would sign a bill including Yosemite Valley in the national park. The trick would be to get the precious land released from the state.

Thus the greatest shift in American environmental policy took place not in the nation's capital, but in the wild Sierra setting of this book. As Frederick Turner aptly concludes, "Two major figures in American history enacted in microcosm one of the culture's most persistent dreams: creative truancy in the wild heart of the New World."

The scenario would be far more improbable today. A bearded mountain man who tried to sneak off alone into the wilds with the president of the United States would be lucky to escape alive after ground and air pursuit by armed park rangers and Secret Service agents. If a "secret" forest meeting was scheduled in advance, all spontaneity would be wrung from it, with helicopters arriving days before to check out the site and surround it with proper protection and access, and with network television crews poised on the perimeter like scavengers.

At that time, the Southern Pacific Railroad was the West's largest private landowner, having acquired vast right-of-way tracts from the government. The railroad's president, Edward H. Harriman, was widely considered to be a robber baron because of the company's earlier rapacious history. But Harriman saw the value in preserving forests and wildlands through which trains could bring visitors from all over the world. When John Muir and the Sierra Club tried to get the California state legislature to return Yosemite Valley to the federal government, entrenched private interests in the valley put up a battle. The urbanizing and mismanagement of the valley floor began to escalate into a national scandal, for the exploiters appeared to control the legislature. As a last resort, Muir asked Harriman for help. After the powerfully connected railroad man lobbied behind the scenes, the bill passed the state senate in 1905 by a single vote.

Congress did not immediately vote to include Yosemite Valley in the national park. In fact, in 1905 it first voted to reduce the size of the park, removing some of the most splendid high country around Mount Ritter as well as the Minarets. Yosemite Valley was added in 1906 only after Muir again asked Harriman to pull strings with politicians, this time in Washington. Muir inadvertently furthered the railroad's quest to profit from Yosemite tourism when Congress followed his guidelines in the bill it passed creating the park instead of following a counterproposal for a Yosemite just one-fifth as large.

Muir was able to lean on Harriman because they had become close friends on a boat trip to Alaska six years earlier. Harriman had suffered a major setback to his health and had decided to aim his life away from business and toward creative individuals whose lives revolved around pursuing truth and helping humanity. Muir was one of twenty-three scientists invited to accompany Harriman on this journey, during which, at Muir's request, they veered off course in search of a through passage and happened to discover Glacier Bay. (Also aboard was Harriman's seven-year-old son, Averell, later to become more famous than his father as governor of New York, ambassador to Russia, and secretary of commerce.) Though Harriman was cool and dictatorial in his management style, he opened himself up enough to Muir to be told, in the old mountaineer's forthright manner, that he was the poorer of the two, for Muir knew how much money he needed to live well, and Harriman did not.

Though John Muir became a national household name in his later years, he was reluctant to lock himself inside for months to write books. He had published only one book, *The Mountains of California*, a collection of old magazine articles, fourteen years before. The impetus to write came from Harriman, who was impressed by Muir's ability to tell stories that were like natural spoken literature. In 1908 he invited Muir to his family ranch at Klamath Lake, Oregon, saying, "You come up to the lodge, and I will show you how to write books. The trouble with you is you are too slow in your beginnings. You plan and brood too much. Begin, begin, begin!" At the lodge, Harriman constantly goaded Muir into telling the story of his life while a stenographer followed him around, taking down every word. The result was a surprisingly coherent 1,000-page manuscript that Muir later edited down into *The Story of My Boyhood and Youth*, published as a book in 1913.

The five-year gap between note taking and publication came about because Muir did not begin to edit the notes when Harriman sent them. Instead he decided to work on what he thought was a more significant project, taking out his faded blue forty-year-old notebooks from the summer of

INTRODUCTION

1869. With the resolve that Harriman had fostered, he sat down and edited his notes, with only minor revisions, into *My First Summer in the Sierra,* which was published in 1911.

Critics have remarked on the youthful exuberance and spontaneity of this book compared to the measured perspective of an old man that lurks behind each page of *The Story of My Boyhood and Youth.* Muir's faithful depictions of direct experience, written in the field, are the essence of the book you are about to read.

When I joined a two-week Sierra Club backcountry outing with my family in 1951, my father brought along his worn, dark blue first edition of *My First Summer.* He would read aloud passages that moved him as we looked out over a landscape that had been included in Muir's original Yosemite National Park, only to be dropped in 1905. I enjoyed the poetry of the words but didn't realize their lasting meaning until decades later. I was a typically smug ten-year-old, and my world view was based on simplistic assumptions. The wild meadows and streams would always be here, along with the peaks and the sky. That the wilderness around me could ever have roads, mines, ski lifts, condos, fast-food stands, or "no trespassing" signs never entered my young mind. It seemed only natural that the words my father read described exactly what was here now and that Muir's adventures on foot were an anticipation of my own. My naiveté not only reflected the rampant optimism of the postwar era but also was a harbinger of a generally unrecognized global environmental crisis in the making.

How different my experience has been from that of people who briefly visit Yosemite Valley by automobile and castigate it as a dreadfully urbanized park. Politicians, journalists on short deadlines, and television crews lugging heavy cameras rarely venture beyond the seven-square-mile valley floor, which was overdeveloped even before it was included in the national park. They fail to experience the remaining 1,200 square miles of national park, of which 94 percent is designated wilderness. The high-country meadows are now in better shape than they were when Muir accompanied thousands of domestic sheep as they grazed around Tuolumne Meadows that first summer. In these pages he decries sheep as "hoofed locusts" that ate every leaf for miles around. He never herded sheep again.

Muir would have decried the construction of the highway, straightened in 1957, that cuts through the polished granite and meadows of the Yosemite high country on its way over the crest of the Sierra. And yet Tioga Pass remains the only road that crosses the southern alpine glory of that crest. The John Muir Trail, built after his death

INTRODUCTION

in 1914, winds for 211 roadless miles near the crest through three national parks, one national monument, and two U.S. Forest Service wilderness areas to connect Yosemite with the 14,496-foot summit of Mount Whitney.

Muir's words have led generations of Sierra travelers to follow his example and adapt their behavior to the natural character of the land, departing from the pioneer ethic of adapting land to human needs. This ethic, quite new at the time, explains why the High Sierra has not been developed as the Alps have, with hut systems, roads, and gondolas breaking the mountain fastness. It gave birth to a view of the world that has become synonymous with the modern environmental movement.

John Muir's path to a new consciousness began in his early years, when he spent extensive time in the wilds without trying to change his surroundings to fit his needs. He first lived, then espoused, a revolutionary ethic of minimal human impact upon the natural state of the earth. Some historians have devalued his contribution to environmental philosophy by showing that his ideas often echoed the earlier writings of Henry David Thoreau and Ralph Waldo Emerson. Yet Muir clearly moved beyond the more passive transcendental thought of these great men from the East Coast. The copy of *The Prose Works of Ralph*

INTRODUCTION

Waldo Emerson that he wore out by carrying it all around the Sierra in his pack, annotating the margins with his own comments, was an 1870 first edition — published a year after Muir wrote his notes for *My First Summer.*

Muir was greatly disappointed when Emerson declined an invitation to join him in the High Sierra during a visit to Yosemite in 1871. He later wrote Emerson that the incident was "a sad commentary on culture and the glorious transcendentalism." In reply, Emerson made the audacious request that Muir "bring to an early close your absolute contracts with any yet unvisited glaciers" and go east as his permanent guest. Muir refused and continued actively to explore wildness with both body and mind throughout his life. Holding this physical connection was essential to his later greatness, both as a writer and as a conservationist. Without that sense of wholeness, his 1869 field notes might never have been published as *My First Summer in the Sierra.*

GALEN ROWELL
Berkeley, California
February 1998

MY FIRST SUMMER
IN THE SIERRA

CHAPTER I

THROUGH THE FOOTHILLS WITH A FLOCK OF SHEEP

In the great Central Valley of California there are only two seasons — spring and summer. The spring begins with the first rainstorm, which usually falls in November. In a few months the wonderful flowery vegetation is in full bloom, and by the end of May it is dead and dry and crisp, as if every plant had been roasted in an oven.

Then the lolling, panting flocks and herds are driven to the high, cool, green pastures of the Sierra. I was longing for the mountains about this time, but money was scarce and I could n't see how a bread supply was to be kept up. While I was anxiously brooding on the bread problem, so troublesome to wanderers, and trying to believe that I might learn to live like the wild animals, gleaning nourishment here and there from seeds, berries, etc., sauntering and climbing in joyful independence

3

of money or baggage, Mr. Delaney, a sheep-owner, for whom I had worked a few weeks, called on me, and offered to engage me to go with his shepherd and flock to the headwaters of the Merced and Tuolumne Rivers — the very region I had most in mind. I was in the mood to accept work of any kind that would take me into the mountains whose treasures I had tasted last summer in the Yosemite region. The flock, he explained, would be moved gradually higher through the successive forest belts as the snow melted, stopping for a few weeks at the best places we came to. These I thought would be good centers of observation from which I might be able to make many telling excursions within a radius of eight or ten miles of the camps to learn something of the plants, animals, and rocks; for he assured me that I should be left perfectly free to follow my studies. I judged, however, that I was in no way the right man for the place, and freely explained my shortcomings, confessing that I was wholly unacquainted with the topography of the upper mountains, the streams that would have to be crossed, and the wild sheep-eating animals, etc.; in short that, what with bears, coyotes, rivers, cañons, and thorny, bewildering chaparral, I feared that half or more of his flock would be lost. Fortunately these shortcomings seemed insignificant to Mr. Delaney. The main thing, he said, was to have a man about the camp whom he could trust to see that the shepherd did his duty, and he assured me that the difficulties that seemed so formidable at a distance would vanish as we went on; encouraging me further by saying that the shepherd would do all the herding, that I could study plants and rocks and scenery as much as I liked, and that he would himself accompany us to the first main camp and make occasional visits to our higher ones to replenish our store of provisions and see how we prospered. Therefore I concluded to go, though still fearing, when I saw the silly sheep bouncing one by one through the narrow gate of the home corral to be counted, that of the two thousand and fifty many would never return.

I was fortunate in getting a fine St. Bernard dog for a companion. His master, a hunter with whom I was slightly acquainted, came to me as soon as he heard that I was going to spend the summer in the Sierra and begged me to take his favorite dog, Carlo, with me, for he feared that if he were compelled to stay all summer on the plains the fierce heat might be the death of him. "I think I can trust you to be kind to him," he said, "and I am sure he will be good to you. He knows all about the moun-

5

tain animals, will guard the camp, assist in managing the sheep, and in every way be found able and faithful." Carlo knew we were talking about him, watched our faces, and listened so attentively that I fancied he understood us. Calling him by name, I asked him if he was willing to go with me. He looked me in the face with eyes expressing wonderful intelligence, then turned to his master, and after permission was given by a wave of the hand toward me and a farewell patting caress, he quietly followed me as if he perfectly understood all that had been said and had known me always.

June 3, 1869. This morning provisions, camp-kettles, blankets, plant-press, etc., were packed on two horses, the flock headed for the tawny foothills, and away we sauntered in a cloud of dust: Mr. Delaney, bony and tall, with sharply hacked profile like Don Quixote, leading the pack-horses, Billy, the proud shepherd, a Chinaman and a Digger Indian to assist in driving for the first few days in the brushy foothills, and myself with notebook tied to my belt.

The home ranch from which we set out is on the south side of the Tuolumne River near French Bar, where the foothills of metamorphic gold-bearing slates dip below the stratified deposits of the Central Valley. We had not gone more than a mile before some of the old leaders of the flock showed by the eager, inquiring way they ran and looked ahead that they were thinking of the high pastures they had enjoyed last summer. Soon the whole flock seemed to be hopefully excited, the mothers calling their lambs, the lambs replying in tones wonderfully human, their fondly quavering calls interrupted now and then by hastily snatched mouthfuls of withered grass. Amid all this seeming babel of baas as they streamed over the hills every mother and child recognized each other's voice. In case a tired lamb, half asleep in the smothering dust, should fail to answer, its mother would come running back through the flock toward the spot whence its last response was heard, and refused to be comforted until she found it, the one of a thousand, though to our eyes and ears all seemed alike.

The flock traveled at the rate of about a mile an hour, outspread in the form of an irregular triangle, about a hundred yards wide at the base, and a hundred and fifty yards long, with a crooked, ever-changing point made up of the strongest foragers, called the "leaders," which, with the most active of those scattered along the ragged sides of the "main body," hastily explored nooks in the rocks and bushes for grass and leaves; the lambs and feeble old mothers dawdling in the rear were called the "tail end."

About noon the heat was hard to bear; the poor sheep panted pitifully and tried to stop in the shade of every tree they came to, while we gazed with eager longing through the dim burning glare toward the snowy mountains and streams, though not one was in sight. The landscape is only wavering foothills roughened here and there with bushes and trees and outcropping masses of slate. The trees, mostly the blue oak (*Quercus Douglasii*), are about thirty to forty feet high, with pale blue-green leaves and white bark, sparsely planted on the thinnest soil or in crevices of rocks beyond the reach of grass fires. The slates in many places rise abruptly through the tawny grass in sharp lichen-covered slabs like tombstones in deserted burying-grounds. With the exception of the oak and four or five species of manzanita and ceanothus, the vegetation of the foothills is mostly the same as that of the plains. I saw this region in the early spring, when it was a charming landscape garden full of birds and bees and flowers. Now the scorching weather makes everything dreary. The ground is full of cracks, lizards glide about on the rocks, and ants in amazing numbers, whose tiny sparks of life only burn the brighter with the heat,

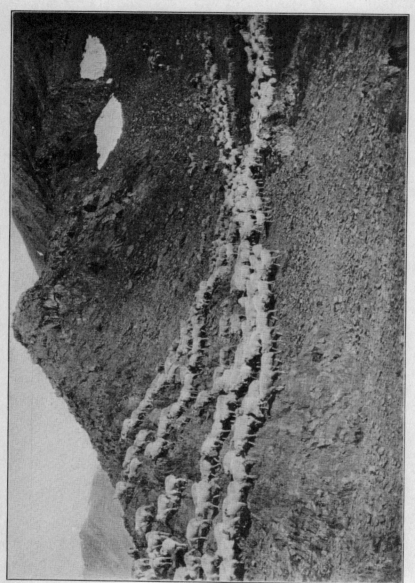

SHEEP IN THE MOUNTAINS

fairly quiver with unquenchable energy as they run in long lines to fight and gather food. How it comes that they do not dry to a crisp in a few seconds' exposure to such sun-fire is marvelous. A few rattlesnakes lie coiled in out-of-the-way places, but are seldom seen. Magpies and crows, usually so noisy, are silent now, standing in mixed flocks on the ground beneath the best shade trees, with bills wide open and wings drooped, too breathless to speak; the quails also are trying to keep in the shade about the few tepid alkaline water-holes; cottontail rabbits are running from shade to shade among the ceanothus brush, and occasionally the long-eared hare is seen cantering gracefully across the wider openings.

After a short noon rest in a grove, the poor dust-choked flock was again driven ahead over the brushy hills, but the dim roadway we had been following faded away just where it was most needed, compelling us to stop to look about us and get our bearings. The Chinaman seemed to think we were lost, and chattered in pidgin English concerning the abundance of "litty stick" (chaparral), while the Indian silently scanned the billowy ridges and gulches for openings. Pushing through the thorny jungle, we at length discovered a road trending toward Coulterville, which we followed until an hour before sunset, when we reached a dry ranch and camped for the night.

Camping in the foothills with a flock of sheep is simple and easy, but far from pleasant. The sheep were allowed to pick what they could find in the neighborhood until after sunset, watched by the shepherd, while the others gathered wood, made a fire, cooked, unpacked and fed the horses, etc. About dusk the weary sheep were gathered on the highest open spot near camp, where they willingly bunched close together, and after each mother had found her lamb and suckled it, all lay down and required no attention until morning.

Supper was announced by the call, "Grub!" Each with a tin plate helped himself direct from the pots and pans while chatting about such camp studies as sheep-feed, mines, coyotes, bears, or adventures during the memorable gold days of pay dirt. The Indian kept in the background, saying never a word, as if he belonged to another species. The meal finished, the dogs were fed, the smokers smoked by the fire, and under the influences of fullness and tobacco the calm that settled on their faces seemed almost divine, something like the mellow meditative glow portrayed on the countenances of saints. Then suddenly, as if awakening from a dream, each with a sigh or a grunt

10

knocked the ashes out of his pipe, yawned, gazed at the fire a few moments, said, "Well, I believe I'll turn in," and straightway vanished beneath his blankets. The fire smouldered and flickered an hour or two longer; the stars shone brighter; coons, coyotes, and owls stirred the silence here and there, while crickets and hylas made a cheerful, continuous music, so fitting and full that it seemed a part of the very body of the night. The only discordance came from a snoring sleeper, and the coughing sheep with dust in their throats. In the starlight the flock looked like a big gray blanket.

June 4. The camp was astir at daybreak; coffee, bacon, and beans formed the breakfast, followed by quick dish-washing and packing. A general bleating began about sunrise. As soon as a mother ewe arose, her lamb came bounding and bunting for its breakfast, and after the thousand youngsters had been suckled the flock began to nibble and spread. The restless wethers with ravenous appetites were the first to move, but dared not go far from the main body. Billy and the Indian and the Chinaman kept them headed along the weary road, and allowed them to pick up what little they could find on a breadth of about a quarter of a mile. But as several flocks had already gone ahead of us, scarce a leaf, green or dry, was

11

left; therefore the starving flock had to be hurried on over the bare, hot hills to the nearest of the green pastures, about twenty or thirty miles from here.

The pack-animals were led by Don Quixote, a heavy rifle over his shoulder intended for bears and wolves. This day has been as hot and dusty as the first, leading over gently sloping brown hills, with mostly the same vegetation, excepting the strange-looking Sabine pine (*Pinus Sabiniana*), which here forms small groves or is scattered among the blue oaks. The trunk divides at a height of fifteen or twenty feet into two or more stems, out-leaning or nearly upright, with many straggling branches and long gray needles, casting but little shade. In general appearance this tree looks more like a palm than a pine. The cones are about six or seven inches long, about five in diameter, very heavy, and last long after they fall, so that the ground beneath the trees is covered with them. They make fine resiny, light-giving camp-fires, next to ears of Indian corn the most beautiful fuel I've ever seen. The nuts, the Don tells me, are gathered in large quantities by the Digger Indians for food. They are about as large and hard-shelled as hazelnuts — food and fire fit for the gods from the same fruit.

June 5. This morning a few hours after setting out with the crawling sheep-cloud, we gained the summit of the first well-defined bench on the mountain-flank at Pino Blanco. The Sabine pines interest me greatly. They are so airy and strangely palm-like I was eager to sketch them, and was in a fever of excitement without accomplishing much. I managed to halt long enough, however, to make a tolerably fair sketch of Pino Blanco peak from the southwest side, where there is a small field and vineyard irrigated by a stream that makes a pretty fall on its way down a gorge by the roadside.

After gaining the open summit of this first bench, feeling the natural exhilaration due to the slight elevation of a thousand feet or so, and the hopes excited concerning the outlook to be obtained, a magnificent section of the Merced Valley at what is called Horse-shoe Bend came full in sight — a glorious wilderness that seemed to be calling with a thousand songful voices. Bold, down-sweeping slopes, feathered with pines and clumps of manzanita with sunny, open spaces between them, make up most of the foreground; the middle and background present fold beyond fold of finely modeled hills and ridges rising into mountain-like masses in the distance, all covered with a shaggy growth of chaparral, mostly adenostoma, planted so marvelously close and even that it looks like soft, rich plush without a single tree or bare spot. As far as the eye can reach it extends, a heaving, swelling sea of green as regular and continuous as that produced by the heaths of Scotland. The sculpture of the landscape is as striking in its main lines as in its lavish richness of detail; a grand congregation of massive heights with the river shining between, each carved into smooth, graceful folds without leaving a single rocky angle exposed, as if the delicate fluting and ridging fashioned out of metamorphic slates had been carefully sandpapered. The whole landscape showed design, like man's noblest sculptures. How wonderful the power of its beauty! Gazing awestricken, I might have left everything for it. Glad, endless work would then be mine tracing the forces that have brought forth its features, its rocks and plants and animals and glorious weather. Beauty beyond thought everywhere, beneath, above, made and being made forever. I gazed and gazed and longed and admired until the dusty sheep and packs were far out of sight, made hurried notes and a sketch, though there was no need of either, for the colors and lines and expression of this di-

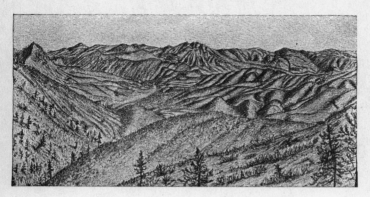

HORSESHOE BEND, MERCED RIVER

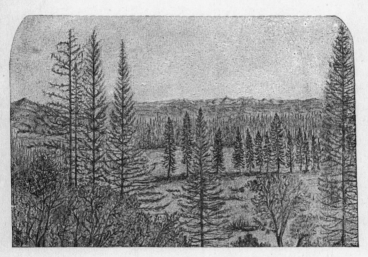

ON SECOND BENCH. EDGE OF THE MAIN FOREST
BELT, ABOVE COULTERVILLE, NEAR
GREELEY'S MILL

vine landscape-countenance are so burned into mind and heart they surely can never grow dim.

The evening of this charmed day is cool, calm, cloudless, and full of a kind of lightning I have never seen before — white glowing cloud-shaped masses down among the trees and bushes, like quick-throbbing fireflies in the Wisconsin meadows rather than the so-called "wild fire." The spreading hairs of the horses' tails and sparks from our blankets show how highly charged the air is.

June 6. We are now on what may be called the second bench or plateau of the Range, after making many small ups and downs over belts of hill-waves, with, of course, corresponding changes in the vegetation. In open spots many of the lowland compositæ are still to be found, and some of the Mariposa tulips and other conspicuous members of the lily family; but the characteristic blue oak of the foothills is left below, and its place is taken by a fine large species (*Quercus Californica*) with deeply lobed deciduous leaves, picturesquely divided trunk, and broad, massy, finely lobed and modeled head. Here also at a height of about twenty-five hundred feet we come to the edge of the great coniferous forest, made up mostly of yellow pine with just a few sugar pines. We are now in the mountains and they are in us, kindling enthusiasm, making every nerve quiver, filling every pore and cell of us. Our flesh-and-bone tabernacle seems transparent as glass to the beauty about us, as if truly an inseparable part of it, thrilling with the air and trees, streams and rocks, in the waves of the sun, — a part of all nature, neither old nor young, sick nor well, but immortal. Just now I can hardly conceive of any bodily condition dependent on food or breath any more than the ground or the sky. How glorious a conversion, so complete and wholesome it is, scarce memory enough of old bondage days left as a standpoint to view it from! In this newness of life we seem to have been so always.

Through a meadow opening in the pine woods I see snowy peaks about the headwaters of the Merced above Yosemite. How near they seem and how clear their outlines on the blue air, or rather *in* the blue air; for they seem to be saturated with it. How consuming strong the invitation they extend! Shall I be allowed to go to them? Night and day I'll pray that I may, but it seems too good to be true. Some one worthy will go, able for the Godful work, yet as far as I can I must drift about these love-monument mountains, glad to be a servant of servants in so holy a wilderness.

Found a lovely lily (*Calochortus albus*) in a shady adenostoma thicket near Coulterville, in company with *Adiantum Chilense*. It is white with a faint purplish tinge inside at the base of the petals, a most impressive plant, pure as a snow crystal, one of the plant saints that all must love and be made so much the purer by it every time it is seen. It puts the roughest mountaineer on his good behavior. With this plant the whole world would seem rich though none other existed. It is not easy to keep on with the camp cloud while such plant people are standing preaching by the wayside.

During the afternoon we passed a fine meadow bounded by stately pines, mostly the arrowy yellow pine, with here and there a noble sugar pine, its feathery arms outspread above the spires of its companion species in marked contrast; a glorious tree, its cones fifteen to twenty inches long, swinging like tassels at the ends of the branches with superb ornamental effect. Saw some logs of this species at the Greeley Mill. They are round and regular as if turned in a lathe, excepting the butt cuts, which have a few buttressing projections. The fragrance of the sugary sap is delicious and scents the mill and lumber yard. How beautiful the ground beneath this pine thickly strewn with slender needles and grand cones, and the piles of cone-scales, seed-wings and shells around the instep of each tree where the squirrels have been feasting! They get the seeds by cutting off the scales at the base in regular order, following their spiral arrangement, and the two seeds at the base of each scale, a hundred or two in a cone, must make a good meal. The yellow pine cones and those of most other species and genera are held upside down on the ground by the Douglas squirrel, and turned around gradually until stripped, while he sits usually with his back to a tree, probably for safety. Strange to say, he never seems to get himself smeared with gum, not even his paws or whiskers — and how cleanly and beautiful in color the cone-litter kitchen-middens he makes.

We are now approaching the region of clouds and cool streams. Magnificent white cumuli appeared about noon above the Yosemite region, — floating fountains refreshing the glorious wilderness, — sky mountains in whose pearly hills and dales the streams take their rise, — blessing with cooling shadows and rain. No rock landscape is more varied in sculpture, none more delicately modeled than these landscapes of the sky;

domes and peaks rising, swelling, white as finest marble and firmly outlined, a most impressive manifestation of world building. Every rain-cloud, however fleeting, leaves its mark, not only on trees and flowers whose pulses are quickened, and on the replenished streams and lakes, but also on the rocks are its marks engraved whether we can see them or not.

I have been examining the curious and influential shrub *Adenostoma fasciculata*, first noticed about Horseshoe Bend. It is very abundant on the lower slopes of the second plateau near Coulterville, forming a dense, almost impenetrable growth that looks dark in the distance. It belongs to the rose family, is about six or eight feet high, has small white flowers in racemes eight to twelve inches long, round needle-like leaves, and reddish bark that becomes shreddy when old. It grows on sun-beaten slopes, and like grass is often swept away by running fires, but is quickly renewed from the roots. Any trees that may have established themselves in its midst are at length killed by these fires, and this no doubt is the secret of the unbroken character of its broad belts. A few manzanitas, which also rise again from the root after consuming fires, make out to dwell with it, also a few bush compositæ — baccharis and linosyris, and some liliaceous plants, mostly calochortus and brodiæa, with deepset bulbs safe from fire. A multitude of birds and "wee, sleekit, cow'rin', tim'rous beasties" find good homes in its deepest thickets, and the open bays and lanes that fringe the margins of its main belts offer shelter and food to the deer when winter storms drive them down from their high mountain pastures. A most admirable plant! It is now in bloom, and I like to wear its pretty fragrant racemes in my buttonhole.

Azalea occidentalis, another charming shrub, grows beside cool streams hereabouts and much higher in the Yosemite region. We found it this evening in bloom a few miles above Greeley's Mill, where we are camped for the night. It is closely related to the rhododendrons, is very showy and fragrant, and everybody must like it not only for itself but for the shady alders and willows, ferny meadows, and living water associated with it.

Another conifer was met to-day — incense cedar (*Libocedrus decurrens*), a large tree with warm yellow-green foliage in flat plumes like those of arborvitæ, bark cinnamon-colored, and as the boles of the old trees are without limbs they make striking pillars in the woods where the sun chances to shine on them — a

worthy companion of the kingly sugar and
yellow pines. I feel strangely attracted to this
tree. The brown close-grained wood, as well
as the small scale-like leaves, is fragrant, and
the flat over-lapping plumes make fine beds,
and must shed the rain well. It would be delightful to be storm-bound beneath one of
these noble, hospitable, inviting old trees, its
broad sheltering arms bent down like a tent,
incense rising from the fire made from its dry
fallen branches, and a hearty wind chanting
overhead. But the weather is calm to-night,
and our camp is only a sheep camp. We are
near the North Fork of the Merced. The
night wind is telling the wonders of the upper
mountains, their snow fountains and gardens,
forests and groves; even their topography is
in its tones. And the stars, the everlasting
sky lilies, how bright they are now that we
have climbed above the lowland dust! The
horizon is bounded and adorned by a spiry
wall of pines, every tree harmoniously related
to every other; definite symbols, divine hiero-
glyphics written with sunbeams. Would I
could understand them! The stream flowing
past the camp through ferns and lilies and
alders makes sweet music to the ear, but the
pines marshaled around the edge of the sky
make a yet sweeter music to the eye. Divine

beauty all. Here I could stay tethered forever with just bread and water, nor would I
be lonely; loved friends and neighbors, as love
for everything increased, would seem all the
nearer however many the miles and mountains between us.

June 7. The sheep were sick last night, and
many of them are still far from well, hardly
able to leave camp, coughing, groaning, looking wretched and pitiful, all from eating the
leaves of the blessed azalea. So at least say
the shepherd and the Don. Having had but
little grass since they left the plains, they are
starving, and so eat anything green they can
get. "Sheep-men" call azalea "sheep-poison,"
and wonder what the Creator was thinking
about when he made it — so desperately does
sheep business blind and degrade, though
supposed to have a refining influence in the
good old days we read of. The California sheep
owner is in haste to get rich, and often does,
now that pasturage costs nothing, while the
climate is so favorable that no winter food
supply, shelter-pens, or barns are required.
Therefore large flocks may be kept at slight
expense, and large profits realized, the money
invested doubling, it is claimed, every other
year. This quickly acquired wealth usually
creates desire for more. Then indeed the wool

is drawn close down over the poor fellow's eyes, dimming or shutting out almost everything worth seeing.

As for the shepherd, his case is still worse, especially in winter when he lives alone in a cabin. For, though stimulated at times by hopes of one day owning a flock and getting rich like his boss, he at the same time is likely to be degraded by the life he leads, and seldom reaches the dignity or advantage — or disadvantage — of ownership. The degradation in his case has for cause one not far to seek. He is solitary most of the year, and solitude to most people seems hard to bear. He seldom has much good mental work or recreation in the way of books. Coming into his dingy hovel-cabin at night, stupidly weary, he finds nothing to balance and level his life with the universe. No, after his dull drag all day after the sheep, he must get his supper; he is likely to slight this task and try to satisfy his hunger with whatever comes handy. Perhaps no bread is baked; then he just makes a few grimy flapjacks in his unwashed frying-pan, boils a handful of tea, and perhaps fries a few strips of rusty bacon. Usually there are dried peaches or apples in the cabin, but he hates to be bothered with the cooking of them, just swallows the bacon and flapjacks, and depends on the genial stupefaction of tobacco for the rest. Then to bed, often without removing the clothing worn during the day. Of course his health suffers, reacting on his mind; and seeing nobody for weeks or months, he finally becomes semi-insane or wholly so.

The shepherd in Scotland seldom thinks of being anything but a shepherd. He has probably descended from a race of shepherds and inherited a love and aptitude for the business almost as marked as that of his collie. He has but a small flock to look after, sees his family and neighbors, has time for reading in fine weather, and often carries books to the fields with which he may converse with kings. The oriental shepherd, we read, called his sheep by name; they knew his voice and followed him. The flocks must have been small and easily managed, allowing piping on the hills and ample leisure for reading and thinking. But whatever the blessings of sheep-culture in other times and countries, the California shepherd, as far as I've seen or heard, is never quite sane for any considerable time. Of all Nature's voices baa is about all he hears. Even the howls and ki-yis of coyotes might be blessings if well heard, but he hears them only through a blur of mutton and wool, and they do him no good.

THROUGH THE FOOTHILLS

The sick sheep are getting well, and the shepherd is discoursing on the various poisons lurking in these high pastures — azalea, kalmia, alkali. After crossing the North Fork of the Merced we turned to the left toward Pilot Peak, and made a considerable ascent on a rocky, brush-covered ridge to Brown's Flat, where for the first time since leaving the plains the flock is enjoying plenty of green grass. Mr. Delaney intends to seek a permanent camp somewhere in the neighborhood, to last several weeks.

Before noon we passed Bower Cave, a delightful marble palace, not dark and dripping, but filled with sunshine, which pours into it through its wide-open mouth facing the south. It has a fine, deep, clear little lake with mossy banks embowered with broad-leaved maples, all under ground, wholly unlike anything I have seen in the cave line even in Kentucky, where a large part of the State is honeycombed with caves. This curious specimen of subterranean scenery is located on a belt of marble that is said to extend from the north end of the Range to the extreme south. Many other caves occur on the belt, but none like this, as far as I have learned, combining as it does sunny outdoor brightness and vegetation with the crystalline beauty of the underworld. It is claimed by a Frenchman, who has fenced and locked it, placed a boat on the lakelet and seats on the mossy bank under the maple trees, and charges a dollar admission fee. Being on one of the ways to the Yosemite Valley, a good many tourists visit it during the travel months of summer, regarding it as an interesting addition to their Yosemite wonders.

Poison oak or poison ivy (*Rhus diversiloba*), both as a bush and a scrambler up trees and rocks, is common throughout the foothill region up to a height of at least three thousand feet above the sea. It is somewhat troublesome to most travelers, inflaming the skin and eyes, but blends harmoniously with its companion plants, and many a charming flower leans confidingly upon it for protection and shade. I have oftentimes found the curious twining lily (*Stropholirion Californicum*) climbing its branches, showing no fear but rather congenial companionship. Sheep eat it without apparent ill effects; so do horses to some extent, though not fond of it, and to many persons it is harmless. Like most other things not apparently useful to man, it has few friends, and the blind question, "Why was it made?" goes on and on with never a guess that first of all it might have been made for itself.

THROUGH THE FOOTHILLS

Brown's Flat is a shallow fertile valley on the top of the divide between the North Fork of the Merced and Bull Creek, commanding magnificent views in every direction. Here the adventurous pioneer David Brown made his headquarters for many years, dividing his time between gold-hunting and bear-hunting. Where could lonely hunter find a better solitude? Game in the woods, gold in the rocks, health and exhilaration in the air, while the colors and cloud furniture of the sky are ever inspiring through all sorts of weather. Though sternly practical, like most pioneers, old David seems to have been uncommonly fond of scenery. Mr. Delaney, who knew him well, tells me that he dearly loved to climb to the summit of a commanding ridge to gaze abroad over the forest to the snow-clad peaks and sources of the rivers, and over the foreground valleys and gulches to note where miners were at work or claims were abandoned, judging by smoke from cabins and camp-fires, the sounds of axes, etc.; and when a rifle-shot was heard, to guess who was the hunter, whether Indian or some poacher on his wide domain. His dog Sandy accompanied him everywhere, and well the little hairy mountaineer knew and loved his master and his master's aims. In deer-hunting he had but little to do, trotting behind his master as he slowly made his way through the wood, careful not to step heavily on dry twigs, scanning open spots in the chaparral, where the game loves to feed in the early morning and towards sunset; peering cautiously over ridges as new outlooks were reached, and along the meadowy borders of streams. But when bears were hunted, little Sandy became more important, and it was as a bear-hunter that Brown became famous. His hunting method, as described by Mr. Delaney, who had passed many a night with him in his lonely cabin and learned his stories, was simply to go slowly and silently through the best bear pastures, with his dog and rifle and a few pounds of flour, until he found a fresh track and then follow it to the death, paying no heed to the time required. Wherever the bear went he followed, led by little Sandy, who had a keen nose and never lost the track, however rocky the ground. When high open points were reached, the likeliest places were carefully scanned. The time of year enabled the hunter to determine approximately where the bear would be found, — in the spring and early summer on open spots about the banks of streams and springy places eating grass and clover and lupines, or in dry meadows feasting on strawberries; toward the end of summer, on

28

dry ridges, feasting on manzanita berries, sitting on his haunches, pulling down the laden branches with his paws, and pressing them together so as to get good compact mouthfuls however much mixed with twigs and leaves; in the Indian summer, beneath the pines, chewing the cones cut off by the squirrels, or occasionally climbing a tree to gnaw and break off the fruitful branches. In late autumn, when acorns are ripe, Bruin's favorite feeding-grounds are groves of the California oak in park-like cañon flats. Always the cunning hunter knew where to look, and seldom came upon Bruin unawares. When the hot scent showed the dangerous game was nigh, a long halt was made, and the intricacies of the topography and vegetation leisurely scanned to catch a glimpse of the shaggy wanderer, or to at least determine where he was most likely to be.

"Whenever," said the hunter, "I saw a bear before it saw me I had no trouble in killing it. I just studied the lay of the land and got to leeward of it no matter how far around I had to go, and then worked up to within a few hundred yards or so, at the foot of a tree that I could easily climb, but too small for the bear to climb. Then I looked well to the condition of my rifle, took off my boots so as to climb well if necessary, and waited until

the bear turned its side in clear view when
I could make a sure or at least a good shot.
In case it showed fight I climbed out of reach.
But bears are slow and awkward with their
eyes, and being to leeward of them they could
not scent me, and I often got in a second shot
before they noticed the smoke. Usually, however, they run when wounded and hide in the
brush. I let them run a good safe time before
I ventured to follow them, and Sandy was
pretty sure to find them dead. If not, he
barked and drew their attention, and occasionally rushed in for a distracting bite, so that
I was able to get to a safe distance for a final
shot. Oh, yes, bear-hunting is safe enough when
followed in a safe way, though like every other
business it has its accidents, and little doggie
and I have had some close calls. Bears like
to keep out of the way of men as a general
thing, but if an old, lean, hungry mother with
cubs met a man on her own ground she would,
in my opinion, try to catch and eat him. This
would be only fair play anyhow, for we eat
them, but nobody hereabout has been used
for bear grub that I know of."

Brown had left his mountain home ere we
arrived, but a considerable number of Digger
Indians still linger in their cedar-bark huts
on the edge of the flat. They were attracted

in the first place by the white hunter whom they had learned to respect, and to whom they looked for guidance and protection against their enemies the Pah Utes, who sometimes made raids across from the east side of the Range to plunder the stores of the comparatively feeble Diggers and steal their wives.

CHAPTER II

June 8. The sheep, now grassy and good-natured, slowly nibbled their way down into the valley of the North Fork of the Merced at the foot of Pilot Peak Ridge to the place selected by the Don for our first central camp, a picturesque hopper-shaped hollow formed by converging hill slopes at a bend of the river. Here racks for dishes and provisions were made in the shade of the river-bank trees, and beds of fern fronds, cedar plumes, and various flowers, each to the taste of its owner, and a corral back on the open flat for the wool.

June 9. How deep our sleep last night in the mountain's heart, beneath the trees and stars, hushed by solemn-sounding waterfalls and many small soothing voices in sweet accord whispering peace! And our first pure mountain day, warm, calm, cloudless, — how immeasurable it seems, how serenely wild! I can scarcely remember its beginning. Along the river, over the hills, in the ground, in the sky, spring work is going on with joyful enthusiasm, new life, new beauty, unfolding, unrolling in glorious exuberant extravagance,

32

— new birds in their nests, new winged creatures in the air, and new leaves, new flowers, spreading, shining, rejoicing everywhere.

The trees about the camp stand close, giving ample shade for ferns and lilies, while back from the bank most of the sunshine reaches the ground, calling up the grasses and flowers in glorious array, tall bromus waving like bamboos, starry compositæ, monardella, Mariposa tulips, lupines, gilias, violets, glad children of light. Soon every fern frond will be unrolled, great beds of common pteris and woodwardia along the river, wreaths and rosettes of pellæa and cheilanthes on sunny rocks. Some of the woodwardia fronds are already six feet high.

A handsome little shrub, *Chamœbatia foliolosa*, belonging to the rose family, spreads a yellow-green mantle beneath the sugar pines for miles without a break, not mixed or roughened with other plants. Only here and there a Washington lily may be seen nodding above its even surface, or a bunch or two of tall bromus as if for ornament. This fine carpet shrub begins to appear at, say, twenty-five hundred or three thousand feet above sea level, is about knee high or less, has brown branches, and the largest stems are only about half an inch in diameter. The leaves, light yellow green, thrice pinnate and finely cut, give them a rich ferny appearance, and they are dotted with minute glands that secrete wax with a peculiar pleasant odor that blends finely with the spicy fragrance of the pines. The flowers are white, five eighths of an inch in diameter, and look like those of the strawberry. Am delighted with this little bush. It is the only true carpet shrub of this part of the Sierra. The manzanita, rhamnus, and most of the species of ceanothus make shaggy rugs and border fringes rather than carpets or mantles.

The sheep do not take kindly to their new pastures, perhaps from being too closely hemmed in by the hills. They are never fully at rest. Last night they were frightened, probably by bears or coyotes prowling and planning for a share of the grand mass of mutton.

June 10. Very warm. We get water for the camp from a rock basin at the foot of a picturesque cascading reach of the river where it is well stirred and made lively without being beaten into dusty foam. The rock here is black metamorphic slate, worn into smooth knobs in the stream channels, contrasting with the fine gray and white cascading water as it glides and glances and falls in lace-like sheets and braided overfolding currents. Tufts of sedge growing

34

on the rock knobs that rise above the surface
produce a charming effect, the long elastic
leaves arching over in every direction, the tips
of the longest drooping into the current, which
dividing against the projecting rocks makes
still finer lines, uniting with the sedges to see
how beautiful the happy stream can be made.
Nor is this all, for the giant saxifrage also is
growing on some of the knob rock islets, firmly
anchored and displaying their broad, round,
umbrella-like leaves in showy groups by themselves, or above the sedge tufts. The flowers
of this species (*Saxifraga peltata*) are purple,
and form tall glandular racemes that are in
bloom before the appearance of the leaves. The
fleshy root-stocks grip the rock in cracks and
hollows, and thus enable the plant to hold on
against occasional floods, — a marked species
employed by Nature to make yet more beautiful the most interesting portions of these cool
clear streams. Near camp the trees arch over
from bank to bank, making a leafy tunnel full
of soft subdued light, through which the young
river sings and shines like a happy living creature.

Heard a few peals of thunder from the upper
Sierra, and saw firm white bossy cumuli rising
back of the pines. This was about noon.

June 11. On one of the eastern branches of

the river discovered some charming cascades with a pool at the foot of each of them. White dashing water, a few bushes and tufts of carex on ledges leaning over with fine effect, and large orange lilies assembled in superb groups on fertile soil-beds beside the pools.

There are no large meadows or grassy plains near camp to supply lasting pasture for our thousands of busy nibblers. The main dependence is ceanothus brush on the hills and tufted grass patches here and there, with lupines and pea-vines among the flowers on sunny open spaces. Large areas have already been stripped bare, or nearly so, compelling the poor hungry wool bundles to scatter far and wide, keeping the shepherds and dogs at the top of their speed to hold them within bounds. Mr. Delaney has gone back to the plains, taking the Indian and Chinaman with him, leaving instruction to keep the flock here or hereabouts until his return, which he promised would not be long delayed.

How fine the weather is! Nothing more celestial can I conceive. How gently the winds blow! Scarce can these tranquil air-currents be called winds. They seem the very breath of Nature, whispering peace to every living thing. Down in the camp dell there is no swaying of tree-tops; most of the time not a leaf moves.

I don't remember having seen a single lily swinging on its stalk, though they are so tall the least breeze would rock them. What grand bells these lilies have! Some of them big enough for children's bonnets. I have been sketching them, and would fain draw every leaf of their wide shining whorls and every curved and spotted petal. More beautiful, better kept gardens cannot be imagined. The species is *Lilium pardalinum*, five to six feet high, leaf-whorls a foot wide, flowers about six inches wide, bright orange, purple spotted in the throat, segments revolute — a majestic plant.

June 12. A slight sprinkle of rain — large drops far apart, falling with hearty pat and plash on leaves and stones and into the mouths of the flowers. Cumuli rising to the eastward. How beautiful their pearly bosses! How well they harmonize with the upswelling rocks beneath them. Mountains of the sky, solid-looking, finely sculptured, their richly varied topography wonderfully defined. Never before have I seen clouds so substantial looking in form and texture. Nearly every day toward noon they rise with visible swelling motion as if new worlds were being created. And how fondly they brood and hover over the gardens and forests with their cooling shadows and

showers, keeping every petal and leaf in glad
health and heart. One may fancy the clouds
themselves are plants, springing up in the sky-
fields at the call of the sun, growing in beauty
until they reach their prime, scattering rain
and hail like berries and seeds, then wilting and
dying.

The mountain live oak, common here and a
thousand feet or so higher, is like the live oak
of Florida, not only in general appearance,
foliage, bark, and wide-branching habit, but in
its tough, knotty, unwedgeable wood. Standing alone with plenty of elbow room, the largest
trees are about seven to eight feet in diameter
near the ground, sixty feet high, and as wide or
wider across the head. The leaves are small
and undivided, mostly without teeth or wavy
edging, though on young shoots some are
sharply serrated, both kinds being found on
the same tree. The cups of the medium-sized
acorns are shallow, thick walled, and covered
with a golden dust of minute hairs. Some of
the trees have hardly any main trunk, dividing near the ground into large wide-spreading
limbs, and these, dividing again and again,
terminate in long, drooping, cord-like branchlets, many of which reach nearly to the ground,
while a dense canopy of short, shining, leafy
branchlets forms a round head which looks

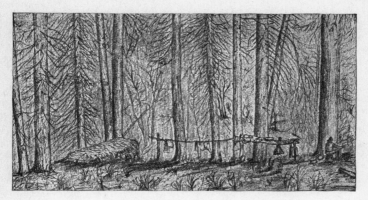

CAMP, NORTH FORK OF THE MERCED

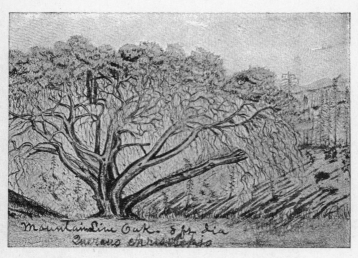

MOUNTAIN LIVE OAK (*Quercus chrysolepis*),
EIGHT FEET IN DIAMETER

something like a cumulus cloud when the sunshine is pouring over it.

A marked plant is the bush poppy (*Dendromecon rigidum*), found on the hot hillsides near camp, the only woody member of the order I have yet met in all my walks. Its flowers are bright orange yellow, an inch to two inches wide, fruit-pods three or four inches long, slender and curving, — height of bushes about four feet, made up of many slim, straight branches, radiating from the root, — a companion of the manzanita and other sun-loving chaparral shrubs.

June 13. Another glorious Sierra day in which one seems to be dissolved and absorbed and sent pulsing onward we know not where. Life seems neither long nor short, and we take no more heed to save time or make haste than do the trees and stars. This is true freedom, a good practical sort of immortality. Yonder rises another white skyland. How sharply the yellow pine spires and the palm-like crowns of the sugar pines are outlined on its smooth white domes. And hark! the grand thunder billows booming, rolling from ridge to ridge, followed by the faithful shower.

A good many herbaceous plants come thus far up the mountains from the plains, and are now in flower, two months later than their lowland relatives. Saw a few columbines to-day. Most of the ferns are in their prime, — rock ferns on the sunny hillsides, cheilanthes, pellæa, gymnogramme; woodwardia, aspidium, woodsia along the stream banks, and the common *Pteris aquilina* on sandy flats. This last, however common, is here making shows of strong, exuberant, abounding beauty to set the botanist wild with admiration. I measured some scarce full grown that are more than seven feet high. Though the commonest and most widely distributed of all the ferns, I might almost say that I never saw it before. The broad-shouldered fronds held high on smooth stout stalks growing close together, overleaning and overlapping, make a complete ceiling, beneath which one may walk erect over several acres without being seen, as if beneath a roof. And how soft and lovely the light streaming through this living ceiling, revealing the arching branching ribs and veins of the fronds as the framework of countless panes of pale green and yellow plant-glass nicely fitted together — a fairyland created out of the commonest fern-stuff.

The smaller animals wander about as if in a tropical forest. I saw the entire flock of sheep vanish at one side of a patch and reappear a hundred yards farther on at the other, their

progress betrayed only by the jerking and
trembling of the fronds; and strange to say
very few of the stout woody stalks were broken.
I sat a long time beneath the tallest fronds,
and never enjoyed anything in the way of a
bower of wild leaves more strangely impressive.
Only spread a fern frond over a man's head and
worldly cares are cast out, and freedom and
beauty and peace come in. The waving of a
pine tree on the top of a mountain, — a magic
wand in Nature's hand, — every devout mountaineer knows its power; but the marvelous
beauty value of what the Scotch call a breckan
in a still dell, what poet has sung this? It
would seem impossible that any one, however
incrusted with care, could escape the Godful
influence of these sacred fern forests. Yet this
very day I saw a shepherd pass through one
of the finest of them without betraying more
feeling than his sheep. "What do you think
of these grand ferns?" I asked. "Oh, they're
only d——d big brakes," he replied.

Lizards of every temper, style, and color
dwell here, seemingly as happy and compan-
ionable as the birds and squirrels. Lowly,
gentle fellow mortals, enjoying God's sunshine,
and doing the best they can in getting a living,
I like to watch them at their work and play.
They bear acquaintance well, and one likes

them the better the longer one looks into their
beautiful, innocent eyes. They are easily
tamed, and one soon learns to love them, as
they dart about on the hot rocks, swift as
dragon-flies. The eye can hardly follow them;
but they never make long-sustained runs, usually only about ten or twelve feet, then a sudden stop, and as sudden a start again; going
all their journeys by quick, jerking impulses.
These many stops I find are necessary as rests,
for they are short-winded, and when pursued
steadily are soon out of breath, pant pitifully,
and are easily caught. Their bodies are more
than half tail, but these tails are well managed,
never heavily dragged nor curved up as if hard
to carry; on the contrary, they seem to follow
the body lightly of their own will. Some are
colored like the sky, bright as bluebirds, others
gray like the lichened rocks on which they hunt
and bask. Even the horned toad of the plains
is a mild, harmless creature, and so are the
snake-like species which glide in curves with
true snake motion, while their small, undevel-
oped limbs drag as useless appendages. One
specimen fourteen inches long which I observed
closely made no use whatever of its tender,
sprouting limbs, but glided with all the soft,
sly ease and grace of a snake. Here comes a
little, gray, dusty fellow who seems to know

and trust me, running about my feet, and looking up cunningly into my face. Carlo is watching, makes a quick pounce on him, for the fun of the thing I suppose; but Liz has shot away from his paws like an arrow, and is safe in the recesses of a clump of chaparral. Gentle saurians, dragons, descendants of an ancient and mighty race, Heaven bless you all and make your virtues known! for few of us know as yet that scales may cover fellow creatures as gentle and lovable as feathers, or hair, or cloth.

Mastodons and elephants used to live here no great geological time ago, as shown by their bones, often discovered by miners in washing gold-gravel. And bears of at least two species are here now, besides the California lion or panther, and wild cats, wolves, foxes, snakes, scorpions, wasps, tarantulas; but one is almost tempted at times to regard a small savage black ant as the master existence of this vast mountain world. These fearless, restless, wandering imps, though only about a quarter of an inch long, are fonder of fighting and biting than any beast I know. They attack every living thing around their homes, often without cause as far as I can see. Their bodies are mostly jaws curved like ice-hooks, and to get work for these weapons seems to be their chief aim and pleasure. Most of their colonies are established in

living oaks somewhat decayed or hollowed, in
which they can conveniently build their cells.
These are chosen probably because of their
strength as opposed to the attacks of animals
and storms. They work both day and night,
creep into dark caves, climb the highest trees,
wander and hunt through cool ravines as well
as on hot, unshaded ridges, and extend their
highways and byways over everything but
water and sky. From the foothills to a mile
above the level of the sea nothing can stir without their knowledge; and alarms are spread in
an incredibly short time, without any howl or
cry that we can hear. I can't understand
the need of their ferocious courage; there seems
to be no common sense in it. Sometimes, no
doubt, they fight in defense of their homes,
but they fight anywhere and always wherever they can find anything to bite. As soon
as a vulnerable spot is discovered on man or
beast, they stand on their heads and sink their
jaws, and though torn limb from limb, they
will yet hold on and die biting deeper. When
I contemplate this fierce creature so widely distributed and strongly intrenched, I see that
much remains to be done ere the world is
brought under the rule of universal peace and
love.

On my way to camp a few minutes ago, I

passed a dead pine nearly ten feet in diameter. It has been enveloped in fire from top to bottom so that now it looks like a grand black pillar set up as a monument. In this noble shaft a colony of large jet-black ants have established themselves, laboriously cutting tunnels and cells through the wood, whether sound or decayed. The entire trunk seems to have been honeycombed, judging by the size of the talus of gnawed chips like sawdust piled up around its base. They are more intelligent looking than their small, belligerent, strong-scented brethren, and have better manners, though quick to fight when required. Their towns are carved in fallen trunks as well as in those left standing, but never in sound, living trees or in the ground. When you happen to sit down to rest or take notes near a colony, some wandering hunter is sure to find you and come cautiously forward to discover the nature of the intruder and what ought to be done. If you are not too near the town and keep perfectly still he may run across your feet a few times, over your legs and hands and face, up your trousers, as if taking your measure and getting comprehensive views, then go in peace without raising an alarm. If, however, a tempting spot is offered or some suspicious movement excites him, a bite follows, and such a bite! I fancy that a bear or wolf bite is not to be compared with it. A quick electric flame of pain flashes along the outraged nerves, and you discover for the first time how great is the capacity for sensation you are possessed of. A shriek, a grab for the animal, and a bewildered stare follow this bite of bites as one comes back to consciousness from sudden eclipse. Fortunately, if careful, one need not be bitten oftener than once or twice in a lifetime. This wonderful electric species is about three fourths of an inch long. Bears are fond of them, and tear and gnaw their home-logs to pieces, and roughly devour the eggs, larvæ, parent ants, and the rotten or sound wood of the cells, all in one spicy acid hash. The Digger Indians also are fond of the larvæ and even of the perfect ants, so I have been told by old mountaineers. They bite off and reject the head, and eat the tickly acid body with keen relish. Thus are the poor biters bitten, like every other biter, big or little, in the world's great family.

There is also a fine, active, intelligent-looking red species, intermediate in size between the above. They dwell in the ground, and build large piles of seed husks, leaves, straw, etc., over their nests. Their food seems to be mostly insects and plant leaves, seeds and sap. How many mouths Nature has to fill, how many neighbors we have, how little we know about them, and how seldom we get in each other's way! Then to think of the infinite numbers of smaller fellow mortals, invisibly small, compared with which the smallest ants are as mastodons.

June 14. The pool-basins below the falls and cascades hereabouts, formed by the heavy down-plunging currents, are kept nicely clean and clear of detritus. The heavier parts of the material swept over the falls are heaped up a short distance in front of the basins in the form of a dam, thus tending, together with erosion, to increase their size. Sudden changes, however, are effected during the spring floods, when the snow is melting and the upper tributaries are roaring loud from "bank to brae." Then boulders that have fallen into the channels, and which the ordinary summer and winter currents were unable to move, are suddenly swept forward as by a mighty besom, hurled over the falls into these pools, and piled up in a new dam together with part of the old one, while some of the smaller boulders are carried farther down stream and variously lodged according to size and shape, all seeking rest where the force of the current is less than the resistance they are able to offer. But the greatest changes made in these relations of fall, pool,

and dam are caused, not by the ordinary spring
floods, but by extraordinary ones that occur
at irregular intervals. The testimony of trees
growing on flood boulder deposits shows that
a century or more has passed since the last
master flood came to awaken everything movable to go swirling and dancing on wonderful
journeys. These floods may occur during the
summer, when heavy thunder-showers, called
"cloud-bursts," fall on wide, steeply inclined
stream basins furrowed by converging channels, which suddenly gather the waters together
into the main trunk in booming torrents of
enormous transporting power, though short
lived.

One of these ancient flood boulders stands
firm in the middle of the stream channel, just
below the lower edge of the pool dam at the
foot of the fall nearest our camp. It is a nearly
cubical mass of granite about eight feet high,
plushed with mosses over the top and down
the sides to ordinary high-water mark. When
I climbed on top of it to-day and lay down to
rest, it seemed the most romantic spot I had
yet found — the one big stone with its mossy
level top and smooth sides standing square and
firm and solitary, like an altar, the fall in front
of it bathing it lightly with the finest of the
spray, just enough to keep its moss cover fresh;

the clear green pool beneath, with its foam-
bells and its half circle of lilies leaning forward
like a band of admirers, and flowering dogwood
and alder trees leaning over all in sun-sifted
arches. How soothingly, restfully cool it is beneath that leafy, translucent ceiling, and how
delightful the water music — the deep bass
tones of the fall, the clashing, ringing spray, and
infinite variety of small low tones of the current
gliding past the side of the boulder-island, and
glinting against a thousand smaller stones
down the ferny channel! All this shut in; every
one of these influences acting at short range
as if in a quiet room. The place seemed holy,
where one might hope to see God.

After dark, when the camp was at rest, I
groped my way back to the altar boulder and
passed the night on it, — above the water,
beneath the leaves and stars, — everything
still more impressive than by day, the fall seen
dimly white, singing Nature's old love song
with solemn enthusiasm, while the stars peering through the leaf-roof seemed to join in the
white water's song. Precious night, precious
day to abide in me forever. Thanks be to God
for this immortal gift.

June 15. Another reviving morning. Down
the long mountain-slopes the sunbeams pour,
gilding the awakening pines, cheering every

needle, filling every living thing with joy. Robins are singing in the alder and maple groves, the same old song that has cheered and sweetened countless seasons over almost all of our blessed continent. In this mountain hollow they seem as much at home as in farmers' orchards. Bullock's oriole and the Louisiana tanager are here also, with many warblers and other little mountain troubadours, most of them now busy about their nests.

Discovered another magnificent specimen of the goldcup oak six feet in diameter, a Douglas spruce seven feet, and a twining lily (*Stropholirion*), with stem eight feet long, and sixty rose-colored flowers.

Sugar pine cones are cylindrical, slightly tapered at the end and rounded at the base. Found one to-day nearly twenty-four inches long and six in diameter, the scales being open. Another specimen nineteen inches long; the average length of full-grown cones on trees favorably situated is nearly eighteen inches. On the lower edge of the belt at a height of about twenty-five hundred feet above the sea they are smaller, say a foot to fifteen inches long, and at a height of seven thousand feet or more near the upper limits of its growth in the Yosemite region they are about the same size. This noble tree is an inexhaustible study and

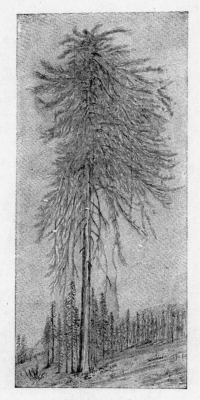

SUGAR PINE

source of pleasure. I never weary of gazing at its grand tassel cones, its perfectly round bole one hundred feet or more without a limb, the fine purplish color of its bark, and its magnificent outsweeping, down-curving feathery arms forming a crown always bold and striking and exhilarating. In habit and general port it looks somewhat like a palm, but no palm that I have yet seen displays such majesty of form and behavior either when poised silent and thoughtful in sunshine, or wide-awake waving in storm winds with every needle quivering. When young it is very straight and regular in form like most other conifers; but at the age of fifty to one hundred years it begins to acquire individuality, so that no two are alike in their prime or old age. Every tree calls for special admiration. I have been making many sketches, and regret that I cannot draw every needle. It is said to reach a height of three hundred feet, though the tallest I have measured falls short of this stature sixty feet or more. The diameter of the largest near the ground is about ten feet, though I've heard of some twelve feet thick or even fifteen. The diameter is held to a great height, the taper being almost imperceptibly gradual. Its companion, the yellow pine, is almost as large. The long silvery foliage of the younger specimens forms

magnificent cylindrical brushes on the top
shoots and the ends of the upturned branches,
and when the wind sways the needles all one
way at a certain angle every tree becomes a
tower of white quivering sun-fire. Well may
this shining species be called the silver pine. The
needles are sometimes more than a foot long,
almost as long as those of the long-leaf pine of
Florida. But though in size the yellow pine almost equals the sugar pine, and in rugged enduring strength seems to surpass it, it is far less
marked in general habit and expression, with
its regular conventional spire and its comparatively small cones clustered stiffly among the
needles. Were there no sugar pine, then would
this be the king of the world's eighty or ninety
species, the brightest of the bright, waving,
worshiping multitude. Were they mere mechanical sculptures, what noble objects they
would still be! How much more throbbing,
thrilling, overflowing, full of life in every fiber
and cell, grand glowing silver-rods — the very
gods of the plant kingdom, living their sublime
century lives in sight of Heaven, watched and
loved and admired from generation to generation! And how many other radiant resiny sun
trees are here and higher up, — libocedrus,
Douglas spruce, silver fir, sequoia. How rich
our inheritance in these blessed mountains,

the tree pastures into which our eyes are turned!

Now comes sundown. The west is all a glory of color transfiguring everything. Far up the Pilot Peak Ridge the radiant host of trees stand hushed and thoughtful, receiving the Sun's good-night, as solemn and impressive a leave-taking as if sun and trees were to meet no more. The daylight fades, the color spell is broken, and the forest breathes free in the night breeze beneath the stars.

June 16. One of the Indians from Brown's Flat got right into the middle of the camp this morning, unobserved. I was seated on a stone, looking over my notes and sketches, and happening to look up, was startled to see him standing grim and silent within a few steps of me, as motionless and weather-stained as an old tree-stump that had stood there for centuries. All Indians seem to have learned this wonderful way of walking unseen, — making themselves invisible like certain spiders I have been observing here, which, in case of alarm, caused, for example, by a bird alighting on the bush their webs are spread upon, immediately bounce themselves up and down on their elastic threads so rapidly that only a blur is visible. The wild Indian power of escaping observation, even where there is little or no cover to hide in, was probably slowly acquired in hard hunting and fighting lessons while trying to approach game, take enemies by surprise, or get safely away when compelled to retreat. And this experience transmitted through many generations seems at length to have become what is vaguely called instinct.

How smooth and changeless seems the surface of the mountains about us! Scarce a track is to be found beyond the range of the sheep except on small open spots on the sides of the streams, or where the forest carpets are thin or wanting. On the smoothest of these open strips and patches deer tracks may be seen, and the great suggestive footprints of bears, which, with those of the many small animals, are scarce enough to answer as a kind of light ornamental stitching or embroidery. Along the main ridges and larger branches of the river Indian trails may be traced, but they are not nearly as distinct as one would expect to find them. How many centuries Indians have roamed these woods nobody knows, probably a great many, extending far beyond the time that Columbus touched our shores, and it seems strange that heavier marks have not been made. Indians walk softly and hurt the landscape hardly more than the birds and squirrels, and their brush and bark huts last hardly longer than those of wood rats, while their more enduring monuments, excepting those wrought on the forests by the fires they made to improve their hunting grounds, vanish in a few centuries.

How different are most of those of the white man, especially on the lower gold region — roads blasted in the solid rock, wild streams dammed and tamed and turned out of their channels and led along the sides of cañons and valleys to work in mines like slaves. Crossing from ridge to ridge, high in the air, on long straddling trestles as if flowing on stilts, or down and up across valleys and hills, imprisoned in iron pipes to strike and wash away hills and miles of the skin of the mountain's face, riddling, stripping every gold gully and flat. These are the white man's marks made in a few feverish years, to say nothing of mills, fields, villages, scattered hundreds of miles along the flank of the Range. Long will it be ere these marks are effaced, though Nature is doing what she can, replanting, gardening, sweeping away old dams and flumes, leveling gravel and boulder piles, patiently trying to heal every raw scar. The main gold storm is over. Calm enough are the gray old miners scratching a bare living in waste diggings here and there. Thundering underground blasting is still going on to feed the pounding quartz mills, but their influence on the landscape is light as compared with that of the pick-and-shovel storms waged a few years ago. Fortunately for Sierra scenery the gold-bearing slates are mostly restricted to the foothills. The region about our camp is still wild, and higher lies the snow about as trackless as the sky.

Only a few hills and domes of cloudland were built yesterday and none at all to-day. The light is peculiarly white and thin, though pleasantly warm. The serenity of this mountain weather in the spring, just when Nature's pulses are beating highest, is one of its greatest charms. There is only a moderate breeze from the summits of the Range at night, and a slight breathing from the sea and the lowland hills and plains during the day, or stillness so complete no leaf stirs. The trees hereabouts have but little wind history to tell.

Sheep, like people, are ungovernable when hungry. Excepting my guarded lily gardens, almost every leaf that these hoofed locusts can reach within a radius of a mile or two from camp has been devoured. Even the bushes are stripped bare, and in spite of dogs and shepherds the sheep scatter to all points of the compass and vanish in dust. I fear some are lost, for one of the sixteen black ones is missing.

June 17. Counted the wool bundles this morning as they bounced through the narrow corral gate. About three hundred are missing, and as the shepherd could not go to seek them, I had to go. I tied a crust of bread to my belt, and with Carlo set out for the upper slopes of the Pilot Peak Ridge, and had a good day, notwithstanding the care of seeking the silly runaways. I went out for wool, and did not come back shorn. A peculiar light circled around the horizon, white and thin like that often seen over the auroral corona, blending into the blue of the upper sky. The only clouds were a few faint flossy pencilings like combed silk. I pushed direct to the boundary of the usual range of the flock, and around it until I found the outgoing trail of the wanderers. It led far up the ridge into an open place surrounded by a hedge-like growth of ceanothus chaparral. Carlo knew what I was about, and eagerly followed the scent until we came up to them, huddled in a timid, silent bunch. They had evidently been here all night and all the forenoon, afraid to go out to feed. Having escaped restraint, they were, like some people we know of, afraid of their freedom, did not know what to do with it, and seemed glad to get back into the old familiar bondage.

June 18. Another inspiring morning, nothing better in any world can be conceived. No description of Heaven that I have ever heard or read of seems half so fine. At noon the clouds occupied about .05 of the sky, white filmy touches drawn delicately on the azure.

The high ridges and hilltops beyond the woolly locusts are now gay with monardella, clarkia, coreopsis, and tall tufted grasses, some of them tall enough to wave like pines. The lupines, of which there are many ill-defined species, are now mostly out of flower, and many of the compositæ are beginning to fade, their radiant corollas vanishing in fluffy pappus like stars in mist.

We had another visitor from Brown's Flat to-day, an old Indian woman with a basket on her back. Like our first caller from the village, she got fairly into camp and was standing in plain view when discovered. How long she had been quietly looking on, I cannot say. Even the dogs failed to notice her stealthy approach. She was on her way, I suppose, to some wild garden, probably for lupine and starchy saxifrage leaves and rootstocks. Her dress was calico rags, far from clean. In every way she seemed sadly unlike Nature's neat well-dressed animals, though living like them on the bounty of the wilderness. Strange that mankind alone is dirty. Had she been clad in fur, or cloth woven of grass or shreddy bark, like the juniper and libocedrus mats, she might then have seemed a rightful part of the wilderness; like a good wolf at least, or bear. But from no point of view that I have found are such debased fellow beings a whit more natural than the glaring tailored tourists we saw that frightened the birds and squirrels.

June 19. Pure sunshine all day. How beautiful a rock is made by leaf shadows! Those of the live oak are particularly clear and distinct, and beyond all art in grace and delicacy, now still as if painted on stone, now gliding softly as if afraid of noise, now dancing, waltzing in swift, merry swirls, or jumping on and off sunny rocks in quick dashes like wave embroidery on seashore cliffs. How true and substantial is this shadow beauty, and with what sublime extravagance is beauty thus multiplied! The big orange lilies are now arrayed in all their glory of leaf and flower. Noble plants, in perfect health, Nature's darlings.

June 20. Some of the silly sheep got caught fast in a tangle of chaparral this morning, like flies in a spider's web, and had to be helped out. Carlo found them and tried to drive them from the trap by the easiest way. How far above sheep are intelligent dogs! No friend and helper can be more affectionate and constant than Carlo. The noble St. Bernard is an honor to his race.

The air is distinctly fragrant with balsam and resin and mint, — every breath of it a gift we may well thank God for. Who could ever guess that so rough a wilderness should yet be so fine, so full of good things. One seems to be in a majestic domed pavilion in which a grand play is being acted with scenery and music and incense, — all the furniture and action so interesting we are in no danger of being called on to endure one dull moment. God himself seems to be always doing his best here, working like a man in a glow of enthusiasm.

June 21. Sauntered along the river-bank to my lily gardens. The perfection of beauty in these lilies of the wilderness is a never-ending source of admiration and wonder. Their rhizomes are set in black mould accumulated in hollows of the metamorphic slates beside the pools, where they are well watered without being subjected to flood action. Every leaf in the level whorls around the tall polished stalks is as finely finished as the petals, and the light and heat required are measured for them and tempered in passing through the branches of over-leaning trees. However strong the winds from the noon rainstorms, they are securely sheltered. Beautiful hypnum carpets bordered with ferns are spread beneath them, violets too, and a few daisies. Everything around them sweet and fresh like themselves.

Cloudland to-day is only a solitary white mountain; but it is so enriched with sunshine and shade, the tones of color on its big domed head and bossy outbulging ridges, and in the hollows and ravines between them, are ineffably fine.

June 22. Unusually cloudy. Besides the periodical shower-bearing cumuli there is a thin, diffused, fog-like cloud overhead. About .75 in all.

June 23. Oh, these vast, calm, measureless mountain days, inciting at once to work and rest! Days in whose light everything seems equally divine, opening a thousand windows to show us God. Nevermore, however weary, should one faint by the way who gains the blessings of one mountain day; whatever his fate, long life, short life, stormy or calm, he is rich forever.

June 24. Our regular allowance of clouds and thunder. Shepherd Billy is in a peck of trouble about the sheep; he declares that they are possessed with more of the evil one than any other flock from the beginning of the

invention of mutton and wool to the last batch
of it. No matter how many are missing, he
will not, he says, go a step to seek them, because, as he reasons, while getting back one
wanderer he would probably lose ten. Therefore runaway hunting must be Carlo's and
mine. Billy's little dog Jack is also giving
trouble by leaving camp every night to visit
his neighbors up the mountain at Brown's
Flat. He is a common-looking cur of no particular breed, but tremendously enterprising
in love and war. He has cut all the ropes and
leather straps he has been tied with, until
his master in desperation, after climbing the
brushy mountain again and again to drag him
back, fastened him with a pole attached to his
collar under his chin at one end, and to a stout
sapling at the other. But the pole gave good
leverage, and by constant twisting during the
night, the fastening at the sapling end was
chafed off, and he set out on his usual journey, dragging the pole through the brush, and
reached the Indian settlement in safety. His
master followed, and making no allowance,
gave him a beating, and swore in bad terms
that next evening he would "fix that infatu-
ated pup" by anchoring him unmercifully to
the heavy cast-iron lid of our Dutch oven,
weighing about as much as the dog. It was

linked directly to his collar close up under the chin, so that the poor fellow seemed unable to stir. He stood quite discouraged until after dark, unable to look about him, or even to lie down unless he stretched himself out with his front feet across the lid, and his head close down between his paws. Before morning, however, Jack was heard far up the height howling Excelsior, cast-iron anchor to the contrary notwithstanding. He must have walked, or rather climbed, erect on his hind legs, clasping the heavy lid like a shield against his breast, a formidable ironclad condition in which to meet his rivals. Next night, dog, pot-lid, and all, were tied up in an old bean-sack, and thus at last angry Billy gained the victory. Just before leaving home, Jack was bitten in the lower jaw by a rattlesnake, and for a week or so his head and neck were swollen to more than double the normal size; nevertheless he ran about as brisk and lively as ever, and is now completely recovered. The only treatment he got was fresh milk — a gallon or two at a time forcibly poured down his sore, poisoned throat.

June 25. Though only a sheep-camp, this grand mountain hollow is home, sweet home, every day growing sweeter, and I shall be sorry to leave it. The lily gardens are safe as yet from the trampling flock. Poor, dusty, raggedy, famishing creatures, I heartily pity them. Many a mile they must go every day to gather their fifteen or twenty tons of chaparral and grass.

June 26. Nuttall's flowering dogwood makes a fine show when in bloom. The whole tree is then snowy white. The involucres are six to eight inches wide. Along the streams it is a good-sized tree thirty to fifty feet high, with a broad head when not crowded by companions. Its showy involucres attract a crowd of moths, butterflies, and other winged people about it for their own, and, I suppose, the tree's advantage. It likes plenty of cool water, and is a great drinker like the alder, willow, and cottonwood, and flourishes best on stream banks, though it often wanders far from streams in damp shady glens beneath the pines, where it is much smaller. When the leaves ripen in the fall, they become more beautiful than the flowers, displaying charming tones of red, purple, and lavender. Another species grows in abundance as a chaparral shrub on the shady sides of the hills, probably *Cornus sessilis*. The leaves are eaten by the sheep. — Heard a few lightning strokes in the distance, with rumbling, mumbling reverberations.

June 27. The beaked hazel (*Corylus rostrata,* var. *Californica*) is common on cool slopes up toward the summit of the Pilot Peak Ridge. There is something peculiarly attractive in the hazel, like the oaks and heaths of the cool countries of our forefathers, and through them our love for these plants has, I suppose, been transmitted. This species is four or five feet high, leaves soft and hairy, grateful to the touch, and the delicious nuts are eagerly gathered by Indians and squirrels. The sky as usual adorned with white noon clouds.

June 28. Warm, mellow summer. The glowing sunbeams make every nerve tingle. The new needles of the pines and firs are nearly full grown and shine gloriously. Lizards are glinting about on the hot rocks; some that live near the camp are more than half tame. They seem attentive to every movement on our part, as if curious to simply look on without suspicion of harm, turning their heads to look back, and making a variety of pretty gestures. Gentle, guileless creatures with beautiful eyes, I shall be sorry to leave them when we leave camp.

June 29. I have been making the acquaintance of a very interesting little bird that flits about the falls and rapids of the main branches of the river. It is not a water-bird in structure, though it gets its living in the water, and never leaves the streams. It is not web-footed, yet it dives fearlessly into deep swirling rapids, evidently to feed at the bottom, using its wings to swim with under water just as ducks and loons do. Sometimes it wades about in shallow places, thrusting its head under from time to time in a jerking, nodding, frisky way that is sure to attract attention. It is about the size of a robin, has short crisp wings serviceable for flying either in water or air, and a tail of moderate size slanted upward, giving it, with its nodding, bobbing manners, a wrennish look. Its color is plain bluish ash, with a tinge of brown on the head and shoulders. It flies from fall to fall, rapid to rapid, with a solid whir of wing-beats like those of a quail, follows the windings of the stream, and usually alights on some rock jutting up out of the current, or on some stranded snag, or rarely on the dry limb of an overhanging tree, perching like regular tree birds when it suits its convenience. It has the oddest, daintiest mincing manners imaginable; and the little fellow can sing too, a sweet, thrushy, fluty song, rather low, not the least boisterous, and much less keen and accentuated than from its vigorous briskness one would be led to look for. What a romantic life this little bird leads on the most beautiful portions of the streams, in a genial climate with shade and cool water and spray to temper the summer heat. No wonder it is a fine singer, considering the stream songs it hears day and night. Every breath the little poet draws is part of a song, for all the air about the rapids and falls is beaten into music, and its first lessons must begin before it is born by the thrilling and quivering of the eggs in unison with the tones of the falls. I have not yet found its nest, but it must be near the streams, for it never leaves them.

June 30. Half cloudy, half sunny, clouds lustrous white. The tall pines crowded along the top of the Pilot Peak Ridge look like six-inch miniatures exquisitely outlined on the satiny sky. Average cloudiness for the day about .25. No rain. And so this memorable month ends, a stream of beauty unmeasured, no more to be sectioned off by almanac arithmetic than sun-radiance or the currents of seas and rivers — a peaceful, joyful stream of beauty. Every morning, arising from the death of sleep, the happy plants and all our fellow animal creatures great and small, and even the rocks, seemed to be shouting, "Awake, awake, rejoice, rejoice, come love us and join in our song. Come! Come!" Looking back through the stillness and romantic enchanting beauty and peace of the camp grove, this June seems the greatest of all the months of my life, the most truly, divinely free, boundless like eternity, immortal. Everything in it seems equally divine — one smooth, pure, wild glow of Heaven's love, never to be blotted or blurred by anything past or to come.

July 1. Summer is ripe. Flocks of seeds are already out of their cups and pods seeking their predestined places. Some will strike root and grow up beside their parents, others flying on the wings of the wind far from them, among strangers. Most of the young birds are full feathered and out of their nests, though still looked after by both father and mother, protected and fed and to some extent educated. How beautiful the home life of birds! No wonder we all love them.

I like to watch the squirrels. There are two species here, the large California gray and the Douglas. The latter is the brightest of all the squirrels I have ever seen, a hot spark of life, making every tree tingle with his prickly toes, a condensed nugget of fresh mountain vigor and valor, as free from disease as a sunbeam. One cannot think of such an animal ever being weary or sick. He seems to think the mountains belong to him, and at first tried

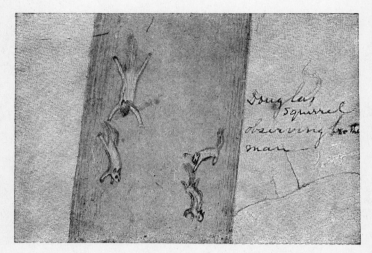

DOUGLAS SQUIRREL OBSERVING BROTHER MAN

to drive away the whole flock of sheep as well
as the shepherd and dogs. How he scolds, and
what faces he makes, all eyes, teeth, and whiskers! If not so comically small, he would indeed be a dreadful fellow. I should like to
know more about his bringing up, his life in
the home knot-hole, as well as in the tree-
tops, throughout all seasons. Strange that I
have not yet found a nest full of young ones.
The Douglas is nearly allied to the red squirrel
of the Atlantic slope, and may have been distributed to this side of the continent by way
of the great unbroken forests of the north.

The California gray is one of the most beautiful, and, next to the Douglas, the most interesting of our hairy neighbors. Compared
with the Douglas he is twice as large, but far
less lively and influential as a worker in the
woods and he manages to make his way
through leaves and branches with less stir
than his small brother. I have never heard
him bark at anything except our dogs. When
in search of food he glides silently from branch
to branch, examining last year's cones, to see
whether some few seeds may not be left between the scales, or gleans fallen ones among
the leaves on the ground, since none of the
present season's crop is yet available. His tail
floats now behind him, now above him, level

or gracefully curled like a wisp of cirrus cloud, every hair in its place, clean and shining and radiant as thistle-down in spite of rough, gummy work. His whole body seems about as unsubstantial as his tail. The little Douglas is fiery, peppery, full of brag and fight and show, with movements so quick and keen they almost sting the onlooker, and the harlequin gyrating show he makes of himself turns one giddy to see. The gray is shy, and oftentimes stealthy in his movements, as if half expecting an enemy in every tree and bush, and back of every log, wishing only to be let alone apparently, and manifesting no desire to be seen or admired or feared. The Indians hunt this species for food, a good cause for caution, not to mention other enemies — hawks, snakes, wild cats. In woods where food is abundant they wear paths through sheltering thickets and over prostrate trees to some favorite pool where in hot and dry weather they drink at nearly the same hour every day. These pools are said to be narrowly watched, especially by the boys, who lie in ambush with bow and arrow, and kill without noise. But, in spite of enemies, squirrels are happy fellows, forest favorites, types of tireless life. Of all Nature's wild beasts, they seem to me the wildest. May we come to know each other better.

NORTH FORK OF THE MERCED

The chaparral-covered hill-slope to the south of the camp, besides furnishing nesting-places for countless merry birds, is the home and hiding-place of the curious wood rat (*Neotoma*), a handsome, interesting animal, always attracting attention wherever seen. It is more like a squirrel than a rat, is much larger, has delicate, thick, soft fur of a bluish slate color, white on the belly; ears large, thin, and translucent; eyes soft, full, and liquid; claws slender, sharp as needles; and as his limbs are strong, he can climb about as well as a squirrel. No rat or squirrel has so innocent a look, is so easily approached, or expresses such confidence in one's good intentions. He seems too fine for the thorny thickets he inhabits, and his hut also is as unlike himself as may be, though softly furnished inside. No other animal inhabitant of these mountains builds houses so large and striking in appearance. The traveler coming suddenly upon a group of them for the first time will not be likely to forget them. They are built of all kinds of sticks, old rotten pieces picked up anywhere, and green prickly twigs bitten from the nearest bushes, the whole mixed with miscellaneous odds and ends of everything movable, such as bits of cloddy earth, stones, bones, deerhorn, etc., piled up in a conical mass as if it were got ready for burning. Some of these curious cabins are six feet high and as wide at the base, and a dozen or more of them are occasionally grouped together, less perhaps for the sake of society than for advantages of food and shelter. Coming through the dense shaggy thickets of some lonely hillside, the solitary explorer happening into one of these strange villages is startled at the sight, and may fancy himself in an Indian settlement, and begin to wonder what kind of reception he is likely to get. But no savage face will he see, perhaps not a single inhabitant, or at most two or three seated on top of their wigwams, looking at the stranger with the mildest of wild eyes, and allowing a near approach. In the center of the rough spiky hut a soft nest is made of the inner fibers of bark chewed to tow, and lined with feathers and the down of various seeds, such as willow and milkweed. The delicate creature in its prickly, thick-walled home suggests a tender flower in a thorny involucre. Some of the nests are built in trees thirty or forty feet from the ground, and even in garrets, as if seeking the company and protection of man, like swallows and linnets, though accustomed to the wildest solitude. Among housekeepers Neotoma has the reputation of a thief, because he carries away everything transportable to his queer hut, — knives, forks, combs,

nails, tin cups, spectacles, etc., — merely, however, to strengthen his fortifications, I guess.
His food at home, as far as I have learned, is
nearly the same as that of the squirrels — nuts,
berries, seeds, and sometimes the bark and
tender shoots of the various species of ceanothus.

July 2. Warm, sunny day, thrilling plant
and animals and rocks alike, making sap and
blood flow fast, and making every particle of
the crystal mountains throb and swirl and
dance in glad accord like star-dust. No dullness anywhere visible or thinkable. No stagna-
tion, no death. Everything kept in joyful
rhythmic motion in the pulses of Nature's big
heart.

Pearl cumuli over the higher mountains —
clouds, not with a silver lining, but all silver.
The brightest, crispest, rockiest-looking clouds,
most varied in features and keenest in outline I
ever saw at any time of year in any country.
The daily building and unbuilding of these
snowy cloud-ranges — the highest Sierra — is
a prime marvel to me, and I gaze at the stupendous white domes, miles high, with ever fresh
admiration. But in the midst of these sky and
mountain affairs a change of diet is pulling us
down. We have been out of bread a few days,
and begin to miss it more than seems reason-

able, for we have plenty of meat and sugar and tea. Strange we should feel food-poor in so rich a wilderness. The Indians put us to shame, so do the squirrels, — starchy roots and seeds and bark in abundance, yet the failure of the meal sack disturbs our bodily balance, and threatens our best enjoyments.

July 3. Warm. Breeze just enough to sift through the woods and waft fragrance from their thousand fountains. The pine and fir cones are growing well, resin and balsam dripping from every tree, and seeds are ripening fast, promising a fine harvest. The squirrels will have bread. They eat all kinds of nuts long before they are ripe, and yet never seem to suffer in stomach.

CHAPTER III

A BREAD FAMINE

July 4. The air beyond the flock range, full of the essences of the woods, is growing sweeter and more fragrant from day to day, like ripening fruit.

Mr. Delaney is expected to arrive soon from the lowlands with a new stock of provisions, and as the flock is to be moved to fresh pastures we shall all be well fed. In the mean time our stock of beans as well as flour has failed — everything but mutton, sugar, and tea. The shepherd is somewhat demoralized, and seems to care but little what becomes of his flock. He says that since the boss has failed to feed him he is not rightly bound to feed the sheep, and swears that no decent white man can climb these steep mountains on mutton alone. "It's not fittin' grub for a white man really white. For dogs and coyotes and Indians it's different. Good grub, good sheep. That's what I say." Such was Billy's Fourth of July oration.

July 5. The clouds of noon on the high Sierra seem yet more marvelously, indescribably beautiful from day to day as one becomes

75

more wakeful to see them. The smoke of the gunpowder burned yesterday on the lowlands, and the eloquence of the orators has probably settled or been blown away by this time. Here every day is a holiday, a jubilee ever sounding with serene enthusiasm, without wear or waste or cloying weariness. Everything rejoicing. Not a single cell or crystal unvisited or forgotten.

July 6. Mr. Delaney has not arrived, and the bread famine is sore. We must eat mutton a while longer, though it seems hard to get accustomed to it. I have heard of Texas pioneers living without bread or anything made from the cereals for months without suffering, using the breast-meat of wild turkeys for bread. Of this kind they had plenty in the good old days when life, though considered less safe, was fussed over the less. The trappers and fur traders of early days in the Rocky Mountain regions lived on bison and beaver meat for months. Salmon-eaters, too, there are among both Indians and whites who seem to suffer little or not at all from the want of bread. Just at this moment mutton seems the least desirable of food, though of good quality. We pick out the leanest bits, and down they go against heavy disgust, causing nausea and an effort to reject the offensive stuff. Tea makes matters worse, if possible. The stomach begins to assert itself as an independent creature with a will of its own. We should boil lupine leaves, clover, starchy petioles, and saxifrage rootstocks like the Indians. We try to ignore our gastric troubles, rise and gaze about us, turn our eyes to the mountains, and climb doggedly up through brush and rocks into the heart of the scenery. A stifled calm comes on, and the day's duties and even enjoyments are languidly got through with. We chew a few leaves of ceanothus by way of luncheon, and smell or chew the spicy monardella for the dull headache and stomach-ache that now lightens, now comes muffling down upon us and into us like fog. At night more mutton, flesh to flesh, down with it, not too much, and there are the stars shining through the cedar plumes and branches above our beds.

July 7. Rather weak and sickish this morning, and all about a piece of bread. Can scarce command attention to my best studies, as if one could n't take a few days' saunter in the Godful woods without maintaining a base on a wheat-field and gristmill. Like caged parrots we want a cracker, any of the hundred kinds — the remainder biscuit of a voyage around the world would answer well enough, nor would the wholesomeness of saleratus biscuit be questioned.

Bread without flesh is a good diet, as on many botanical excursions I have proved. Tea also may easily be ignored. Just bread and water and delightful toil is all I need, — not unreasonably much, yet one ought to be trained and tempered to enjoy life in these brave wilds in full independence of any particular kind of nourishment. That this may be accomplished is manifest, as far as bodily welfare is concerned, in the lives of people of other climes. The Eskimo, for example, gets a living far north of the wheat line, from oily seals and whales. Meat, berries, bitter weeds, and blubber, or only the last, for months at a time; and yet these people all around the frozen shores of our continent are said to be hearty, jolly, stout, and brave. We hear, too, of fish-eaters, carnivorous as spiders, yet well enough as far as stomachs are concerned, while we are so ridiculously helpless, making wry faces over our fare, looking sheepish in digestive distress amid rumbling, grumbling sounds that might well pass for smothered baas. We have a large supply of sugar, and this evening it occurred to me that these belligerent stomachs might possibly, like complaining children, be coaxed with candy. Accordingly the frying-pan was cleansed, and a lot of sugar cooked in it to a sort of wax, but this stuff only made matters worse.

A BREAD FAMINE

Man seems to be the only animal whose food soils him, making necessary much washing and shield-like bibs and napkins. Moles living in the earth and eating slimy worms are yet as clean as seals or fishes, whose lives are one perpetual wash. And, as we have seen, the squirrels in these resiny woods keep themselves clean in some mysterious way; not a hair is sticky, though they handle the gummy cones, and glide about apparently without care. The birds, too, are clean, though they seem to make a good deal of fuss washing and cleaning their feathers. Certain flies and ants I see are in a fix, entangled and sealed up in the sugar-wax we threw away, like some of their ancestors in amber. Our stomachs, like tired muscles, are sore with long squirming. Once I was very hungry in the Bonaventure graveyard near Savannah, Georgia, having fasted for several days; then the empty stomach seemed to chafe in much the same way as now, and a somewhat similar tenderness and aching was produced, hard to bear, though the pain was not acute. We dream of bread, a sure sign we need it. Like the Indians, we ought to know how to get the starch out of fern and saxifrage stalks, lily bulbs, pine bark, etc. Our education has been sadly neglected for many generations. Wild rice would be good. I noticed a leersia in wet meadow edges, but the seeds are small. Acorns are not ripe, nor pine nuts, nor filberts. The inner bark of pine or spruce might be tried. Drank tea until half intoxicated. Man seems to crave a stimulant when anything extraordinary is going on, and this is the only one I use. Billy chews great quantities of tobacco, which I suppose helps to stupefy and moderate his misery. We look and listen for the Don every hour. How beautiful upon the mountains his big feet would be!

In the warm, hospitable Sierra, shepherds and mountain men in general, as far as I have seen, are easily satisfied as to food supplies and bedding. Most of them are heartily content to "rough it," ignoring Nature's fineness as bothersome or unmanly. The shepherd's bed is often only the bare ground and a pair of blankets, with a stone, a piece of wood, or a pack-saddle for a pillow. In choosing the spot, he shows less care than the dogs, for they usually deliberate before making up their minds in so important an affair, going from place to place, scraping away loose sticks and pebbles, and trying for comfort by making many changes, while the shepherd casts himself down anywhere, seemingly the least skilled of all rest seekers. His food, too, even when he has all he wants, is usually far from delicate, either in kind

or cooking. Beans, bread of any sort, bacon,
mutton, dried peaches, and sometimes potatoes
and onions, make up his bill-of-fare, the two
latter articles being regarded as luxuries on
account of their weight as compared with the
nourishment they contain; a half-sack or so of
each may be put into the pack in setting out
from the home ranch and in a few days they
are done. Beans are the main standby, port-
able, wholesome, and capable of going far, besides being easily cooked, although curiously
enough a great deal of mystery is supposed to
lie about the bean-pot. No two cooks quite
agree on the methods of making beans do their
best, and, after petting and coaxing and nurs-
ing the savory mess, — well oiled and mellowed
with bacon boiled into the heart of it, — the
proud cook will ask, after dishing out a quart
or two for trial, "Well, how do you like *my*
beans?" as if by no possibility could they be
like any other beans cooked in the same way,
but must needs possess some special virtue of
which he alone is master. Molasses, sugar, or
pepper may be used to give desired flavors; or
the first water may be poured off and a spoon-
ful or two of ashes or soda added to dissolve or
soften the skins more fully, according to various tastes and notions. But, like casks of wine,
no two potfuls are exactly alike to every palate.

Some are supposed to be spoiled by the moon, by some unlucky day, by the beans having been grown on soil not suitable; or the whole year may be to blame as not favorable for beans.

Coffee, too, has its marvels in the camp kitchen, but not so many, and not so inscrutable as those that beset the bean-pot. A low, complacent grunt follows a mouthful drawn in with a gurgle, and the remark cast forth aimlessly, "That's good coffee." Then another gurgling sip and repetition of the judgment, "*Yes, sir,* that *is* good coffee." As to tea, there are but two kinds, weak and strong, the stronger the better. The only remark heard is, "That tea's weak," otherwise it is good enough and not worth mentioning. If it has been boiled an hour or two or smoked on a pitchy fire, no matter, — who cares for a little tannin or creosote? they make the black beverage all the stronger and more attractive to tobacco-tanned palates.

Sheep-camp bread, like most California camp bread, is baked in Dutch ovens, some of it in the form of yeast powder biscuit, an unwholesome sticky compound leading straight to dyspepsia. The greater part, however, is fermented with sour dough, a handful from each batch being saved and put away in the mouth of the flour sack to inoculate the next.

A BREAD FAMINE

The oven is simply a cast-iron pot, about five inches deep and from twelve to eighteen inches wide. After the batch has been mixed and kneaded in a tin pan the oven is slightly heated and rubbed with a piece of tallow or pork rind. The dough is then placed in it, pressed out against the sides, and left to rise. When ready for baking a shovelful of coals is spread out by the side of the fire and the oven set upon them, while another shovelful is placed on top of the lid, which is raised from time to time to see that the requisite amount of heat is being kept up. With care good bread may be made in this way, though it is liable to be burned or to be sour, or raised too much, and the weight of the oven is a serious objection.

At last Don Delaney comes doon the lang glen — hunger vanishes, we turn our eyes to the mountains, and to-morrow we go climbing toward cloudland.

Never while anything is left of me shall this first camp be forgotten. It has fairly grown into me, not merely as memory pictures, but as part and parcel of mind and body alike. The deep hopper-like hollow, with its majestic trees through which all the wonderful nights the stars poured their beauty. The flowery wildness of the high steep slope toward Brown's Flat, and its bloom-fragrance descending at

the close of the still days. The embowered
river-reaches with their multitude of voices
making melody, the stately flow and rush
and glad exulting onsweeping currents caressing the dipping sedge-leaves and bushes and
mossy stones, swirling in pools, dividing against
little flowery islands, breaking gray and white
here and there, ever rejoicing, yet with deep
solemn undertones recalling the ocean — the
brave little bird ever beside them, singing with
sweet human tones among the waltzing foam-
bells, and like a blessed evangel explaining
God's love. And the Pilot Peak Ridge, its
long withdrawing slopes gracefully modeled
and braided, reaching from climate to climate,
feathered with trees that are the kings of their
race, their ranks nobly marshaled to view,
spire above spire, crown above crown, waving
their long, leafy arms, tossing their cones like
ringing bells — blessed sun-fed mountaineers
rejoicing in their strength, every tree tuneful,
a harp for the winds and the sun. The hazel
and buckthorn pastures of the deer, the sun-
beaten brows purple and yellow with mint and
golden-rods, carpeted with chamœbatia, hum-
ming with bees. And the dawns and sunrises
and sundowns of these mountain days, — the
rose light creeping higher among the stars,
changing to daffodil yellow, the level beams

bursting forth, streaming across the ridges, touching pine after pine, awakening and warming all the mighty host to do gladly their shining day's work. The great sun-gold noons, the alabaster cloud-mountains, the landscape beaming with consciousness like the face of a god. The sunsets, when the trees stood hushed awaiting their good-night blessings. Divine, enduring, unwastable wealth.

CHAPTER IV

TO THE HIGH MOUNTAINS

July 8. Now away we go toward the topmost mountains. Many still, small voices, as well as the noon thunder, are calling, "Come higher." Farewell, blessed dell, woods, gardens, streams, birds, squirrels, lizards, and a thousand others. Farewell. Farewell.

Up through the woods the hoofed locusts streamed beneath a cloud of brown dust. Scarcely were they driven a hundred yards from the old corral ere they seemed to know that at last they were going to new pastures, and rushed wildly ahead, crowding through gaps in the brush, jumping, tumbling like exulting hurrahing flood-waters escaping through a broken dam. A man on each flank kept shouting advice to the leaders, who in their famishing condition were behaving like Gadarene swine; two others drivers were busy with stragglers, helping them out of brush tangles; the Indian, calm, alert, silently watched for wanderers likely to be overlooked; the two dogs ran here and there, at a loss to know what was best to be done, while the Don,

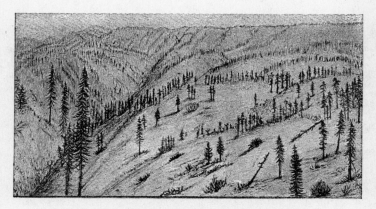

DIVIDE BETWEEN THE TUOLUMNE AND THE
MERCED, BELOW HAZEL GREEN

soon far in the rear, was trying to keep in sight of his troublesome wealth.

As soon as the boundary of the old eaten-out range was passed the hungry horde suddenly became calm, like a mountain stream in a meadow. Thenceforward they were allowed to eat their way as slowly as they wished, care being taken only to keep them headed toward the summit of the Merced and Tuolumne divide. Soon the two thousand flattened paunches were bulged out with sweet-pea vines and grass, and the gaunt, desperate creatures, more like wolves than sheep, became bland and governable, while the howling drivers changed to gentle shepherds, and sauntered in peace.

Toward sundown we reached Hazel Green, a charming spot on the summit of the dividing ridge between the basins of the Merced and Tuolumne, where there is a small brook flowing through hazel and dogwood thickets beneath magnificent silver firs and pines. Here, we are camped for the night, our big fire, heaped high with rosiny logs and branches, is blazing like a sunrise, gladly giving back the light slowly sifted from the sunbeams of centuries of summers; and in the glow of that old sunlight how impressively surrounding objects are brought forward in relief against the outer darkness! Grasses, larkspurs, columbines, lilies, hazel bushes, and the great trees form a circle around the fire like thoughtful spectators, gazing and listening with human-like enthusiasm. The night breeze is cool, for all day we have been climbing into the upper sky, the home of the cloud mountains we so long have admired. How sweet and keen the air! Every breath a blessing. Here the sugar pine reaches its fullest development in size and beauty and number of individuals, filling every swell and hollow and down-plunging ravine almost to the exclusion of other species. A few yellow pines are still to be found as companions, and in the coolest places silver firs; but noble as these are, the sugar pine is king, and spreads long protecting arms above them while they rock and wave in sign of recognition.

We have now reached a height of six thousand feet. In the forenoon we passed along a flat part of the dividing ridge that is planted with manzanita (*Arctostaphylos*), some specimens the largest I have seen. I measured one, the bole of which is four feet in diameter and only eighteen inches high from the ground, where it dissolves into many wide-spreading branches forming a broad round head about ten or twelve feet high, covered with clusters of small narrow-throated pink bells. The leaves are pale green, glandular, and set on edge by a twist of the petiole. The branches seem naked; for the chocolate-colored bark is very smooth and thin, and is shed off in flakes that curl when dry. The wood is red, close-grained, hard, and heavy. I wonder how old these curious tree-bushes are, probably as old as the great pines. Indians and bears and birds and fat grubs feast on the berries, which look like small apples, often rosy on one side, green on the other. The Indians are said to make a kind of beer or cider out of them. There are many species. This one, *Arctostaphylos pungens,* is common hereabouts. No need have they to fear the wind, so low they are and steadfastly rooted. Even the fires that sweep the woods seldom destroy them utterly, for they rise again from the root, and some of the dry ridges they grow on are seldom touched by fire. I must try to know them better.

I miss my river songs to-night. Here Hazel Creek at its topmost springs has a voice like a bird. The wind-tones in the great trees overhead are strangely impressive, all the more because not a leaf stirs below them. But it grows late, and I must to bed. The camp is silent; everybody asleep. It seems extravagant to spend hours so precious in sleep. "He giveth his beloved sleep." Pity the poor beloved needs it, weak, weary, forspent; oh, the pity of it, to sleep in the midst of eternal, beautiful motion instead of gazing forever, like the stars.

July 9. Exhilarated with the mountain air, I feel like shouting this morning with excess of wild animal joy. The Indian lay down away from the fire last night, without blankets, having nothing on, by way of clothing, but a pair of blue overalls and a calico shirt wet with sweat. The night air is chilly at this elevation, and we gave him some horse-blankets, but he did n't seem to care for them. A fine thing to be independent of clothing where it is so hard to carry. When food is scarce, he can live on whatever comes in his way — a few berries, roots, bird eggs, grasshoppers, black ants, fat wasp or bumblebee larvæ, without feeling that he is doing anything worth mention, so I have been told.

Our course to-day was along the broad top of the main ridge to a hollow beyond Crane Flat. It is scarce at all rocky, and is covered with the noblest pines and spruces I have yet seen. Sugar pines from six to eight feet in diameter are not uncommon, with a height of two hundred feet or even more. The silver firs (*Abies concolor* and *A. magnifica*) are ex-

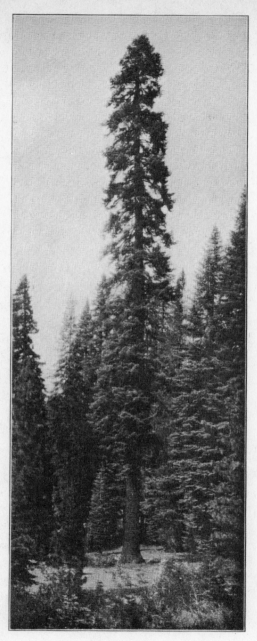

A SILVER FIR, OR RED FIR
(*Abies magnifica*)

ceedingly beautiful, especially the *magnifica*, which becomes more abundant the higher we go. It is of great size, one of the most notable in every way of the giant conifers of the Sierra. I saw specimens that measured seven feet in diameter and over two hundred feet in height, while the average size for what might be called full-grown mature trees can hardly be less than one hundred and eighty or two hundred feet high and five or six feet in diameter; and with these noble dimensions there is a symmetry and perfection of finish not to be seen in any other tree, hereabout at least. The branches are whorled in fives mostly, and stand out from the tall, straight, exquisitely tapered bole in level collars, each branch regularly pinnated like the fronds of ferns, and densely clad with leaves all around the branchlets, thus giving them a singularly rich and sumptuous appearance. The extreme top of the tree is a thick blunt shoot pointing straight to the zenith like an admonishing finger. The cones stand erect like casks on the upper branches. They are about six inches long, three in diameter, blunt, velvety, and cylindrical in form, and very rich and precious looking. The seeds are about three quarters of an inch long, dark reddish brown with brilliant iridescent purple wings, and when ripe, the cone falls to pieces, and the seeds thus set free at a height of one hundred and fifty or two hundred feet have a good send off and may fly considerable distances in a good breeze; and it is when a good breeze is blowing that most of them are shaken free to fly.

The other species, *Abies concolor*, attains nearly as great a height and thickness as the *magnifica*, but the branches do not form such regular whorls, nor are they so exactly pinnated or richly leaf-clad. Instead of growing all around the branchlets, the leaves are mostly arranged in two flat horizontal rows. The cones and seeds are like those of the *magnifica* in form but less than half as large. The bark of the *magnifica* is reddish purple and closely furrowed, that of the *concolor* gray and widely furrowed. A noble pair.

At Crane Flat we climbed a thousand feet or more in a distance of about two miles, the forest growing more dense and the silvery *magnifica* fir forming a still greater portion of the whole. Crane Flat is a meadow with a wide sandy border lying on the top of the divide. It is often visited by blue cranes to rest and feed on their long journeys, hence the name. It is about half a mile long, draining into the Merced, sedgy in the middle, with a margin bright with lilies, columbines, larkspurs, lupines, castilleia, then an outer zone of dry, gently sloping ground starred with a multitude of small flowers — eunanus, mimulus, gilia, with rosettes of spraguea, and tufts of several species of eriogonum and the brilliant zauschneria. The noble forest wall about it is made up of the two silver firs and the yellow and sugar pines, which here seem to reach their highest pitch of beauty and grandeur; for the elevation, six thousand feet or a little more, is not too great for the sugar and yellow pines or too low for the *magnifica* fir, while the *concolor* seems to find this elevation the best possible. About a mile from the north end of the flat there is a grove of *Sequoia gigantea*, the king of all the conifers. Furthermore, the Douglas spruce (*Pseudotsuga Douglasii*) and *Libocedrus decurrens*, and a few two-leaved pines, occur here and there, forming a small part of the forest. Three pines, two silver firs, one Douglas spruce, one sequoia, — all of them, except the two-leaved pine, colossal trees, — are found here together, an assemblage of conifers unrivaled on the globe.

We passed a number of charming garden-like meadows lying on top of the divide or hanging like ribbons down its sides, imbedded in the glorious forest. Some are taken up chiefly with the tall white-flowered *Veratrum Cali-*

fornicum, with boat-shaped leaves about a foot long, eight or ten inches wide, and veined like those of cypripedium, — a robust, hearty, liliaceous plant, fond of water and determined to be seen. Columbine and larkspur grow on the dryer edges of the meadows, with a tall handsome lupine standing waist-deep in long grasses and sedges. Castilleias, too, of several species make a bright show with beds of violets at their feet. But the glory of these forest meadows is a lily (*L. parvum*). The tallest are from seven to eight feet high with magnificent racemes of ten to twenty or more small orange-colored flowers; they stand out free in open ground, with just enough grass and other companion plants about them to fringe their feet, and show them off to best advantage. This is a grand addition to my lily acquaintances, — a true mountaineer, reaching prime vigor and beauty at a height of seven thousand feet or thereabouts. It varies, I find, very much in size even in the same meadow, not only with the soil, but with age. I saw a specimen that had only one flower, and another within a stone's throw had twenty-five. And to think that the sheep should be allowed in these lily meadows! after how many centuries of Nature's care planting and watering them, tucking the bulbs in snugly below winter frost, shading the tender shoots with clouds drawn above them like curtains, pouring refreshing rain, making them perfect in beauty, and keeping them safe by a thousand miracles; yet, strange to say, allowing the trampling of devastating sheep. One might reasonably look for a wall of fire to fence such gardens. So extravagant is Nature with her choicest treasures, spending plant beauty as she spends sunshine, pouring it forth into land and sea, garden and desert. And so the beauty of lilies falls on angels and men, bears and squirrels, wolves and sheep, birds and bees, but as far as I have seen, man alone, and the animals he tames, destroy these gardens. Awkward, lumbering bears, the Don tells me, love to wallow in them in hot weather, and deer with their sharp feet cross them again and again, sauntering and feeding, yet never a lily have I seen spoiled by them. Rather, like gardeners, they seem to cultivate them, pressing and dibbling as required. Anyhow not a leaf or petal seems misplaced.

The trees round about them seem as perfect in beauty and form as the lilies, their boughs whorled like lily leaves in exact order. This evening, as usual, the glow of our campfire is working enchantment on everything within reach of its rays. Lying beneath the firs, it is glorious to see them dipping their spires in the starry sky, the sky like one vast lily meadow in bloom! How can I close my eyes on so precious a night?

July 10. A Douglas squirrel, peppery, pungent autocrat of the woods, is barking overhead this morning, and the small forest birds, so seldom seen when one travels noisily, are out on sunny branches along the edge of the meadow getting warm, taking a sun bath and dew bath — a fine sight. How charming the sprightly confident looks and ways of these little feathered people of the trees! They seem sure of dainty, wholesome breakfasts, and where are so many breakfasts to come from? How helpless should we find ourselves should we try to set a table for them of such buds, seeds, insects, etc., as would keep them in the pure wild health they enjoy! Not a headache or any other ache amongst them, I guess. As for the irrepressible Douglas squirrels, one never thinks of their breakfasts or the possibility of hunger, sickness or death; rather they seem like stars above chance or change, even though we may see them at times busy gathering burrs, working hard for a living.

On through the forest ever higher we go, a cloud of dust dimming the way, thousands of feet trampling leaves and flowers, but in this mighty wilderness they seem but a feeble band, and a thousand gardens will escape their blighting touch. They cannot hurt the trees, though some of the seedlings suffer, and should the woolly locusts be greatly multiplied, as on account of dollar value they are likely to be, then the forests, too, may in time be destroyed. Only the sky will then be safe, though hid from view by dust and smoke, incense of a bad sacrifice. Poor, helpless, hungry sheep, in great part misbegotten, without good right to be, semi-manufactured, made less by God than man, born out of time and place, yet their voices are strangely human and call out one's pity.

Our way is still along the Merced and Tuolumne divide, the streams on our right going to swell the songful Yosemite River, those on our left to the songful Tuolumne, slipping through sunny carex and lily meadows, and breaking into song down a thousand ravines almost as soon as they are born. A more tuneful set of streams surely nowhere exists, or more sparkling crystal pure, now gliding with tinkling whisper, now with merry dimpling rush, in and out through sunshine and shade, shimmering in pools, uniting their currents, bouncing, dancing from form to form over cliffs and inclines, ever more beautiful the farther they go until they pour into the main glacial rivers.

All day I have been gazing in growing admiration at the noble groups of the magnificent silver fir which more and more is taking the ground to itself. The woods above Crane Flat still continue comparatively open, letting in the sunshine on the brown needle-strewn ground. Not only are the individual trees admirable in symmetry and superb in foliage and port, but half a dozen or more often form temple groves in which the trees are so nicely graded in size and position as to seem one. Here, indeed, is the tree-lover's paradise. The dullest eye in the world must surely be quickened by such trees as these.

Fortunately the sheep need little attention, as they are driven slowly and allowed to nip and nibble as they like. Since leaving Hazel Green we have been following the Yosemite trail; visitors to the famous valley coming by way of Coulterville and Chinese Camp pass this way — the two trails uniting at Crane Flat — and enter the valley on the north side. Another trail enters on the south side by way of Mariposa. The tourists we saw were in parties of from three or four to fifteen or twenty, mounted on mules or small mustang ponies. A strange show they made, winding

single file through the solemn woods in gaudy
attire, scaring the wild creatures, and one
might fancy that even the great pines would
be disturbed and groan aghast. But what may
we say of ourselves and the flock?

We are now camped at Tamarack Flat,
within four or five miles of the lower end of
Yosemite. Here is another fine meadow embosomed in the woods, with a deep, clear
stream gliding through it, its banks rounded
and beveled with a thatch of dipping sedges.
The flat is named after the two-leaved pine
(*Pinus contorta*, var. *Murrayana*), common
here, especially around the cool margin of the
meadow. On rocky ground it is a rough, thickset tree, about forty to sixty feet high and one to
three feet in diameter, bark thin and gummy,
branches rather naked, tassels, leaves, and
cones small. But in damp, rich soil it grows
close and slender, and reaches a height at
times of nearly a hundred feet. Specimens
only six inches in diameter at the ground are
often fifty or sixty feet in height, as slender and
sharp in outline as arrows, like the true tamarack (larch) of the Eastern States; hence the
name, though it is a pine.

July 11. The Don has gone ahead on one of
the pack animals to spy out the land to the
north of Yosemite in search of the best point

for a central camp. Much higher than this
we cannot now go, for the upper pastures, said
to be better than any hereabouts, are still
buried in heavy winter snow. Glad I am that
camp is to be fixed in the Yosemite region, for
many a glorious ramble I'll have along the
top of the walls, and then what landscapes I
shall find with their new mountains and
cañons, forests and gardens, lakes and streams
and falls.

We are now about seven thousand feet above
the sea, and the nights are so cool we have to
pile coats and extra clothing on top of our blankets. Tamarack Creek is icy cold, delicious,
exhilarating champagne water. It is flowing
bank-full in the meadow with silent speed, but
only a few hundred yards below our camp the
ground is bare gray granite strewn with boulders, large spaces being without a single tree
or only a small one here and there anchored
in narrow seams and cracks. The boulders,
many of them very large, are not in piles or
scattered like rubbish among loose crumbling
débris as if weathered out of the solid as
boulders of disintegration; they mostly occur
singly, and are lying on a clean pavement on
which the sunshine falls in a glare that contrasts with the shimmer of light and shade we
have been accustomed to in the leafy woods.

And, strange to say, these boulders lying so
still and deserted, with no moving force near
them, no boulder carrier anywhere in sight,
were nevertheless brought from a distance, as
difference in color and composition shows,
quarried and carried and laid down here each
in its place; nor have they stirred, most of
them, through calm and storm since first they
arrived. They look lonely here, strangers in a
strange land, — huge blocks, angular mountain chips, the largest twenty or thirty feet
in diameter, the chips that Nature has made
in modeling her landscapes, fashioning the
forms of her mountains and valleys. And with
what tool were they quarried and carried?
On the pavement we find its marks. The most
resisting unweathered portion of the surface
is scored and striated in a rigidly parallel way,
indicating that the region has been overswept
by a glacier from the northeastward, grinding
down the general mass of the mountains, scoring and polishing, producing a strange, raw,
wiped appearance, and dropping whatever
boulders it chanced to be carrying at the time
it was melted at the close of the Glacial Period.
A fine discovery this. As for the forests we
have been passing through, they are probably
growing on deposits of soil most of which has
been laid down by this same ice agent in the

form of moraines of different sorts, now in great part disintegrated and outspread by post-glacial weathering.

Out of the grassy meadow and down over this ice-planed granite runs the glad young Tamarack Creek, rejoicing, exulting, chanting, dancing in white, glowing, irised falls and cascades on its way to the Merced Cañon, a few miles below Yosemite, falling more than three thousand feet in a distance of about two miles.

All the Merced streams are wonderful singers, and Yosemite is the centre where the main tributaries meet. From a point about half a mile from our camp we can see into the lower end of the famous valley, with its wonderful cliffs and groves, a grand page of mountain manuscript that I would gladly give my life to be able to read. How vast it seems, how short human life when we happen to think of it, and how little we may learn, however hard we try! Yet why bewail our poor inevitable ignorance? Some of the external beauty is always in sight, enough to keep every fibre of us tingling, and this we are able to gloriously enjoy though the methods of its creation may lie beyond our ken. Sing on, brave Tamarack Creek, fresh from your snowy fountains, plash and swirl and dance to your fate in the sea; bathing, cheering every living thing along your way.

Have greatly enjoyed all this huge day, sauntering and seeing, steeping in the mountain influences, sketching, noting, pressing flowers, drinking ozone and Tamarack water. Found the white fragrant Washington lily, the finest of all the Sierra lilies. Its bulbs are buried in shaggy chaparral tangles, I suppose for safety from pawing bears; and its magnificent panicles sway and rock over the top of the rough snow-pressed bushes, while big, bold, blunt-nosed bees drone and mumble in its polleny bells. A lovely flower, worth going hungry and footsore endless miles to see. The whole world seems richer now that I have found this plant in so noble a landscape.

A log house serves to mark a claim to the Tamarack meadow, which may become valuable as a station in case travel to Yosemite should greatly increase. Belated parties occasionally stop here. A white man with an Indian woman is holding possession of the place.

Sauntered up the meadow about sundown, out of sight of camp and sheep and all human mark, into the deep peace of the solemn old woods, everything glowing with Heaven's unquenchable enthusiasm.

July 12. The Don has returned, and again

we go on pilgrimage. "Looking over the Yosemite Creek country," he said, "from the tops of the hills you see nothing but rocks and patches of trees; but when you go down into the rocky desert you find no end of small grassy banks and meadows, and so the country is not half so lean as it looks. There we'll go and stay until the snow is melted from the upper country."

I was glad to hear that the high snow made a stay in the Yosemite region necessary, for I am anxious to see as much of it as possible. What fine times I shall have sketching, studying plants and rocks, and scrambling about the brink of the great valley alone, out of sight and sound of camp!

We saw another party of Yosemite tourists to-day. Somehow most of these travelers seem to care but little for the glorious objects about them, though enough to spend time and money and endure long rides to see the famous valley. And when they are fairly within the mighty walls of the temple and hear the psalms of the falls, they will forget themselves and become devout. Blessed, indeed, should be every pilgrim in these holy mountains!

We moved slowly eastward along the Mono Trail, and early in the afternoon unpacked and camped on the bank of Cascade Creek. The Mono Trail crosses the range by the

Bloody Cañon Pass to gold mines near the north end of Mono Lake. These mines were reported to be rich when first discovered, and a grand rush took place, making a trail necessary. A few small bridges were built over streams where fording was not practicable on account of the softness of the bottom, sections of fallen trees cut out, and lanes made through thickets wide enough to allow the passage of bulky packs; but over the greater part of the way scarce a stone or shovelful of earth has been moved.

The woods we passed through are composed almost wholly of *Abies magnifica*, the companion species, *concolor*, being mostly left behind on account of altitude, while the increasing elevation seems grateful to the charming *magnifica*. No words can do anything like justice to this noble tree. At one place many had fallen during some heavy wind-storm, owing to the loose sandy character of the soil, which offered no secure anchorage. The soil is mostly decomposed and disintegrated moraine material.

The sheep are lying down on a bare rocky spot such as they like, chewing the cud in grassy peace. Cooking is going on, appetites growing keener every day. No lowlander can appreciate the mountain appetite, and the facility with which heavy food called "grub" is disposed of. Eating, walking, resting, seem alike delightful, and one feels inclined to shout lustily on rising in the morning like a crowing cock. Sleep and digestion as clear as the air. Fine spicy plush boughs for bedding we shall have to-night, and a glorious lullaby from this cascading creek. Never was stream more fittingly named, for as far as I have traced it above and below our camp it is one continuous bouncing, dancing, white bloom of cascades. And at the very last unwearied it finishes its wild course in a grand leap of three hundred feet or more to the bottom of the main Yosemite cañon near the fall of Tamarack Creek, a few miles below the foot of the valley. These falls almost rival some of the far-famed Yosemite falls. Never shall I forget these glad cascade songs, the low booming, the roaring, the keen, silvery clashing of the cool water rushing exulting from form to form beneath irised spray; or in the deep still night seen white in the darkness, and its multitude of voices sounding still more impressively sublime. Here I find the little water ouzel as much at home as any linnet in a leafy grove, seeming to take the greater delight the more boisterous the stream. The dizzy precipices, the swift dashing energy displayed, and the thunder tones of the sheer falls are awe-inspiring, but there is nothing awful about this little bird. Its song is sweet and low, and all its gestures, as it flits about amid the loud uproar, bespeak strength and peace and joy. Contemplating these darlings of Nature coming forth from spray-sprinkled nests on the brink of savage streams, Samson's riddle comes to mind, "Out of the strong cometh forth sweetness." A yet finer bloom is this little bird than the foam-bells in eddying pools. Gentle bird, a precious message you bring me. We may miss the meaning of the torrent, but thy sweet voice, only love is in it.

July 13. Our course all day has been eastward over the rim of Yosemite Creek basin and down about halfway to the bottom, where we have encamped on a sheet of glacier-polished granite, a firm foundation for beds. Saw the tracks of a very large bear on the trail, and the Don talked of bears in general. I said I should like to see the maker of these immense tracks as he marched along, and follow him for days, without disturbing him, to learn something of the life of this master beast of the wilderness. Lambs, the Don told me, born in the lowland, that never saw or heard a bear, snort and run in terror when they catch the scent, showing how fully they have inherited a knowledge of their enemy. Hogs, mules, horses, and cattle are afraid of bears, and are seized with ungovernable terror when they approach, particularly hogs and mules. Hogs are frequently driven to pastures in the foothills of the Coast Range and Sierra where acorns are abundant, and are herded in droves of hundreds like sheep. When a bear comes to the range they promptly leave it, emigrating in a body, usually in the night time, the keepers being powerless to prevent; they thus show more sense than sheep, that simply scatter in the rocks and brush and await their fate. Mules flee like the wind with or without riders when they see a bear, and, if picketed, sometimes break their necks in trying to break their ropes, though I have not heard of bears killing mules or horses. Of hogs they are said to be particularly fond, bolting small ones, bones and all, without choice of parts. In particular, Mr. Delaney assured me that all kinds of bears in the Sierra are very shy, and that hunters found far greater difficulty in getting within gunshot of them than of deer or indeed any other animal in the Sierra, and if I was anxious to see much of them I should have to wait and watch with endless Indian patience and pay no attention to anything else.

Night is coming on, the gray rock waves are growing dim in the twilight. How raw and young this region appears! Had the ice sheet that swept over it vanished but yesterday, its traces on the more resisting portions about our camp could hardly be more distinct than they now are. The horses and sheep and all of us, indeed, slipped on the smoothest places.

July 14. How deathlike is sleep in this mountain air, and quick the awakening into newness of life! A calm dawn, yellow and purple, then floods of sun-gold, making everything tingle and glow.

In an hour or two we came to Yosemite Creek, the stream that makes the greatest of all the Yosemite falls. It is about forty feet wide at the Mono Trail crossing, and now about four feet in average depth, flowing about three miles an hour. The distance to the verge of the Yosemite wall, where it makes its tremendous plunge, is only about two miles from here. Calm, beautiful, and nearly silent, it glides with stately gestures, a dense growth of the slender two-leaved pine along its banks, and a fringe of willow, purple spirea, sedges, daisies, lilies, and columbines. Some of the sedges and willow boughs dip into the current, and just outside of the close ranks of trees there is a sunny flat of washed gravelly sand which seems to have been deposited by some ancient flood. It is covered with millions of erethrea, eriogonum, and oxytheca, with more flowers than leaves, forming an even growth, slightly dimpled and ruffled here and there by rosettes of *Spraguea umbellata*. Back of this flowery strip there is a wavy upsloping plain of solid granite, so smoothly ice-polished in many places that it glistens in the sun like glass. In shallow hollows there are patches of trees, mostly the rough form of the two-leaved pine, rather scrawny looking where there is little or no soil. Also a few junipers (*Juniperus occidentalis*), short and stout, with bright cinnamon-colored bark and gray foliage, standing alone mostly, on the sun-beaten pavement, safe from fire, clinging by slight joints, — a sturdy storm-enduring mountaineer of a tree, living on sunshine and snow, maintaining tough health on this diet for perhaps more than a thousand years.

Up towards the head of the basin I see groups of domes rising above the wavelike ridges, and some picturesque castellated masses, and dark strips and patches of silver fir, indicating deposits of fertile soil. Would that I could command the time to study them! What rich excursions one could make in this well-defined basin! Its glacial inscriptions and sculptures, how marvelous they seem, how noble the studies they offer! I tremble with excitement in the dawn of these glorious mountain sublimities, but I can only gaze and wonder, and, like a child, gather here and there a lily, half hoping I may be able to study and learn in years to come.

The drivers and dogs had a lively, laborious time getting the sheep across the creek, the second large stream thus far that they have been compelled to cross without a bridge; the first being the North Fork of the Merced near Bower Cave. Men and dogs, shouting and barking, drove the timid, water-fearing creatures in a close crowd against the bank, but not one of the flock would launch away. While thus jammed, the Don and the shepherd rushed through the frightened crowd to stampede those in front, but this would only cause a break backward, and away they would scamper through the stream-bank trees and scatter over the rocky pavement. Then with the aid of the dogs the runaways would again be gathered and made to face the stream, and again the compacted mass would break away, amid wild shouting and barking that might well have disturbed the stream itself and marred the music of its falls, to which visitors no doubt from all quarters of the globe were listening. "Hold them there! Now hold them there!" shouted the Don; "the front ranks will soon tire of the pressure, and be glad to take to the water, then all will jump in and cross in a hurry." But they did nothing of the kind; they only avoided the pressure by breaking back in scores and hundreds, leaving the beauty of the banks sadly trampled.

If only one could be got to cross over, all would make haste to follow; but that one could not be found. A lamb was caught, carried across, and tied to a bush on the opposite bank, where it cried piteously for its mother. But though greatly concerned, the mother only called it back. That play on maternal affection failed, and we began to fear that we should be forced to make a long roundabout drive and cross the wide-spread tributaries of the creek in succession. This would require several days, but it had its advantages, for I was eager to see the sources of so famous a stream. Don Quixote, however, determined that they must ford just here, and immediately began a sort of siege by cutting down slender pines on the bank and building a corral barely large enough to hold the flock when well pressed together. And as the stream would form one side of the corral he believed that they could easily be forced into the water.

In a few hours the inclosure was completed, and the silly animals were driven in and rammed hard against the brink of the ford.

Then the Don, forcing a way through the compacted mass, pitched a few of the terrified unfortunates into the stream by main strength; but instead of crossing over, they swam about close to the bank, making desperate attempts to get back into the flock. Then a dozen or more were shoved off, and the Don, tall like a crane and a good natural wader, jumped in after them, seized a struggling wether, and dragged it to the opposite shore. But no sooner did he let it go than it jumped into the stream and swam back to its frightened companions in the corral, thus manifesting sheep-nature as unchangeable as gravitation. Pan with his pipes would have had no better luck, I fear. We were now pretty well baffled. The silly creatures would suffer any sort of death rather than cross that stream. Calling a council, the dripping Don declared that starvation was now the only likely scheme to try, and that we might as well camp here in comfort and let the besieged flock grow hungry and cool, and come to their senses, if they had any. In a few minutes after being thus let alone, an adventurer in the foremost rank plunged in and swam bravely to the farther shore. Then suddenly all rushed in pell-mell together, trampling one another under water, while we vainly tried to hold them back. The Don jumped into the

thickest of the gasping, gurgling, drowning
mass, and shoved them right and left as if each
sheep was a piece of floating timber. The current also served to drift them apart; a long bent
column was soon formed, and in a few minutes
all were over and began baaing and feeding as
if nothing out of the common had happened.
That none were drowned seems wonderful. I
fully expected that hundreds would gain the
romantic fate of being swept into Yosemite
over the highest waterfall in the world.

As the day was far spent, we camped a little
way back from the ford, and let the dripping
flock scatter and feed until sundown. The
wool is dry now, and calm, cud-chewing peace
has fallen on all the comfortable band, leaving
no trace of the watery battle. I have seen fish
driven out of the water with less ado than was
made in driving these animals into it. Sheep
brain must surely be poor stuff. Compare to-
day's exhibition with the performances of deer
swimming quietly across broad and rapid rivers,
and from island to island in seas and lakes; or
with dogs, or even with the squirrels that, as
the story goes, cross the Mississippi River on
selected chips, with tails for sails comfortably
trimmed to the breeze. A sheep can hardly be
called an animal; an entire flock is required to
make one foolish individual.

CHAPTER V

THE YOSEMITE

July 15. Followed the Mono Trail up the eastern rim of the basin nearly to its summit, then turned off southward to a small shallow valley that extends to the edge of the Yosemite, which we reached about noon, and encamped. After luncheon I made haste to high ground, and from the top of the ridge on the west side of Indian Cañon gained the noblest view of the summit peaks I have ever yet enjoyed. Nearly all the upper basin of the Merced was displayed, with its sublime domes and cañons, dark up-sweeping forests, and glorious array of white peaks deep in the sky, every feature glowing, radiating beauty that pours into our flesh and bones like heat rays from fire. Sunshine over all; no breath of wind to stir the brooding calm. Never before had I seen so glorious a landscape, so boundless an affluence of sublime mountain beauty. The most extravagant description I might give of this view to any one who has not seen similar landscapes with his own eyes would not so much as hint its grandeur and the spiritual glow that covered it. I shouted and gesticulated in a wild burst of ecstasy, much to the astonishment of St. Bernard Carlo, who came running up to me, manifesting in his intelligent eyes a puzzled concern that was very ludicrous, which had the effect of bringing me to my senses. A brown bear, too, it would seem, had been a spectator of the show I had made of myself, for I had gone but a few yards when I started one from a thicket of brush. He evidently considered me dangerous, for he ran away very fast, tumbling over the tops of the tangled manzanita bushes in his haste. Carlo drew back, with his ears depressed as if afraid, and kept looking me in the face, as if expecting me to pursue and shoot, for he had seen many a bear battle in his day.

Following the ridge, which made a gradual descent to the south, I came at length to the brow of that massive cliff that stands between Indian Cañon and Yosemite Falls, and here the far-famed valley came suddenly into view throughout almost its whole extent. The noble walls — sculptured into endless variety of domes and gables, spires and battlements and plain mural precipices — all a-tremble with the thunder tones of the falling water. The level bottom seemed to be dressed like a garden — sunny meadows here and there, and groves of pine and oak; the river of Mercy sweeping in majesty through the midst of them and flashing back the sunbeams. The great Tissiack, or Half-Dome, rising at the upper end of the valley to a height of nearly a mile, is nobly proportioned and life-like, the most impressive of all the rocks, holding the eye in devout admiration, calling it back again and again from falls or meadows, or even the mountains beyond, — marvelous cliffs, marvelous in sheer dizzy depth and sculpture, types of endurance. Thousands of years have they stood in the sky exposed to rain, snow, frost, earthquake and avalanche, yet they still wear the bloom of youth.

I rambled along the valley rim to the westward; most of it is rounded off on the very brink, so that it is not easy to find places where one may look clear down the face of the wall to the bottom. When such places were found, and I had cautiously set my feet and drawn my body erect, I could not help fearing a little that the rock might split off and let me down, and what a down! — more than three thousand feet. Still my limbs did not tremble, nor did I feel the least uncertainty as to the reliance to be placed on them. My only fear was that a flake of the granite, which in some places showed joints more or less open and running parallel with the face of the cliff, might give way. After withdrawing from such places, excited with the view I had got, I would say to myself, "Now don't go out on the verge again." But in the face of Yosemite scenery cautious remonstrance is vain; under its spell one's body seems to go where it likes with a will over which we seem to have scarce any control.

After a mile or so of this memorable cliff work I approached Yosemite Creek, admiring its easy, graceful, confident gestures as it comes bravely forward in its narrow channel, singing the last of its mountain songs on its way to its fate — a few rods more over the shining granite, then down half a mile in showy foam to another world, to be lost in the Merced, where climate, vegetation, inhabitants, all are different. Emerging from its last gorge, it glides in wide lace-like rapids down a smooth incline into a pool where it seems to rest and compose its gray, agitated waters before taking the grand plunge, then slowly slipping over the lip of the pool basin, it descends another glossy slope with rapidly accelerated speed to the brink of the tremendous cliff, and with sublime, fateful confidence springs out free in the air.

I took off my shoes and stockings and worked my way cautiously down alongside the rushing flood, keeping my feet and hands pressed firmly on the polished rock. The booming, roaring water, rushing past close to my head, was very exciting. I had expected that the sloping apron would terminate with the perpendicular wall of the valley, and that from the foot of it, where it is less steeply inclined, I should be able to lean far enough out to see the forms and behavior of the fall all the way down to the bottom. But I found that there was yet another small brow over which I could not see, and which appeared to be too steep for mortal feet. Scanning it keenly, I discovered a narrow shelf about three inches wide on the very brink, just wide enough for a rest for one's heels. But there seemed to be no way of reaching it over so steep a brow. At length, after careful scrutiny of the surface, I found an irregular edge of a flake of the rock some distance back from the margin of the torrent. If I was to get down to the brink at all that rough edge, which might offer slight finger-holds, was the only way. But the slope beside it looked dangerously smooth and steep, and the swift roaring flood beneath, overhead, and beside me was very nerve-trying. I therefore concluded not to venture farther, but did nevertheless. Tufts of artemisia were growing in clefts of the rock near by, and I filled my mouth with the bitter leaves, hoping they might help to prevent giddiness. Then, with a caution not known in ordinary circumstances, I crept down safely to the little ledge, got my heels well planted on it, then shuffled in a horizontal direction twenty or thirty feet until close to the outplunging current, which, by the time it had descended thus far, was already white. Here I obtained a perfectly free view down into the heart of the snowy, chanting throng of comet-like streamers, into which the body of the fall soon separates.

While perched on that narrow niche I was not distinctly conscious of danger. The tremendous grandeur of the fall in form and sound and motion, acting at close range, smothered the sense of fear, and in such places one's body takes keen care for safety on its own account. How long I remained down there, or how I returned, I can hardly tell. Anyhow I had a glorious time, and got back to camp about dark, enjoying triumphant exhilaration soon followed by dull weariness. Hereafter I'll try to keep from such extravagant, nerve-straining places. Yet such a day is well worth venturing for. My first view of the High Sierra, first view looking down into Yosemite, the death song of Yosemite Creek, and its flight over the vast cliff, each one of these is of itself enough for a great life-long landscape fortune — a most memorable day of days — enjoyment enough to kill if that were possible.

July 16. My enjoyments yesterday afternoon, especially at the head of the fall, were too great for good sleep. Kept starting up last night in a nervous tremor, half awake, fancying that the foundation of the mountain we were camped on had given way and was falling into Yosemite Valley. In vain I roused myself to make a new beginning for sound sleep. The nerve strain had been too great, and again and again I dreamed I was rushing through the air above a glorious avalanche of water and rocks. One time, springing to my feet, I said, "This time it is real — all must die, and where could mountaineer find a more glorious death!"

Left camp soon after sunrise for an all-day ramble eastward. Crossed the head of Indian Basin, forested with *Abies magnifica*, underbrush mostly *Ceanothus cordulatus* and manzanita, a mixture not easily trampled over or penetrated, for the ceanothus is thorny and grows in dense snow-pressed masses, and the manzanita has exceedingly crooked, stubborn branches. From the head of the cañon continued on past North Dome into the basin of Dome or Porcupine Creek. Here are many fine meadows imbedded in the woods, gay with *Lilium parvum* and its companions; the elevation, about eight thousand feet, seems to be best suited for it — saw specimens that were a foot or two higher than my head. Had more magnificent views of the upper mountains, and of the great South Dome, said to be the grandest rock in the world. Well it may be, since it is of such noble dimensions and sculpture. A wonderfully impressive monument, its lines exquisite in fineness, and though sublime in size, is finished like the finest work of art, and seems to be alive.

July 17. A new camp was made to-day in a magnificent silver fir grove at the head of a small stream that flows into Yosemite by way of Indian Cañon. Here we intend to stay several weeks, — a fine location from which to make excursions about the great valley and its fountains. Glorious days I'll have sketching, pressing plants, studying the wonderful topography and the wild animals, our happy fellow mortals and neighbors. But the vast mountains in the distance, shall I ever know them, shall I be allowed to enter into their midst and dwell with them?

We were pelted about noon by a short, heavy rainstorm, sublime thunder reverberating among the mountains and cañons, — some strokes near, crashing, ringing in the tense crisp air with startling keenness, while the distant peaks loomed gloriously through the cloud fringes and sheets of rain. Now the

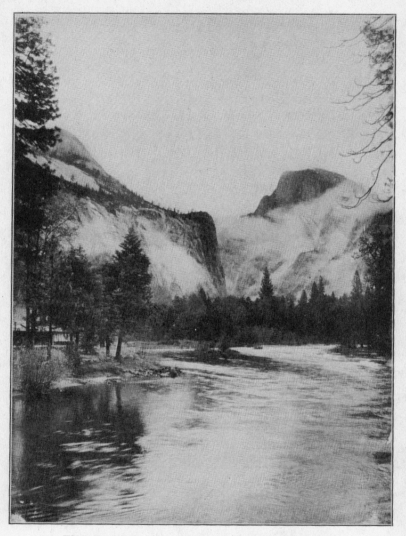

THE NORTH AND SOUTH DOMES, YOSEMITE
NATIONAL PARK

storm is past, and the fresh washed air is full of the essences of the flower gardens and groves. Winter storms in Yosemite must be glorious. May I see them!

Have got my bed made in our new camp, — plushy, sumptuous, and deliciously fragrant, most of it *magnifica* fir plumes, of course, with a variety of sweet flowers in the pillow. Hope to sleep to-night without tottering nerve-dreams. Watched a deer eating ceanothus leaves and twigs.

July 18. Slept pretty well; the valley walls did not seem to fall, though I still fancied myself at the brink, alongside the white, plunging flood, especially when half asleep. Strange the danger of that adventure should be more troublesome now that I am in the bosom of the peaceful woods, a mile or more from the fall, than it was while I was on the brink of it.

Bears seem to be common here, judging by their tracks. About noon we had another rainstorm with keen startling thunder, the metallic, ringing, clashing, clanging notes gradually fading into low bass rolling and muttering in the distance. For a few minutes the rain came in a grand torrent like a waterfall, then hail; some of the hailstones an inch in diameter, hard, icy, and irregular in form, like those oftentimes seen in Wisconsin. Carlo watched them with intelligent astonishment as they came pelting and thrashing through the quivering branches of the trees. The cloud scenery sublime. Afternoon calm, sunful, and clear, with delicious freshness and fragrance from the firs and flowers and steaming ground.

July 19. Watching the daybreak and sunrise. The pale rose and purple sky changing softly to daffodil yellow and white, sunbeams pouring through the passes between the peaks and over the Yosemite domes, making their edges burn; the silver firs in the middle ground catching the glow on their spiry tops, and our camp grove fills and thrills with the glorious light. Everything awakening alert and joyful; the birds begin to stir and innumerable insect people. Deer quietly withdraw into leafy hiding-places in the chaparral; the dew vanishes, flowers spread their petals, every pulse beats high, every life cell rejoices, the very rocks seem to thrill with life. The whole landscape glows like a human face in a glory of enthusiasm, and the blue sky, pale around the horizon, bends peacefully down over all like one vast flower.

About noon, as usual, big bossy cumuli began to grow above the forest, and the rainstorm pouring from them is the most imposing I have yet seen. The silvery zigzag lightning lances are longer than usual, and the thunder gloriously impressive, keen, crashing, intensely concentrated, speaking with such tremendous energy it would seem that an entire mountain is being shattered at every stroke, but probably only a few trees are being shattered, many of which I have seen on my walks hereabouts strewing the ground. At last the clear ringing strokes are succeeded by deep low tones that grow gradually fainter as they roll afar into the recesses of the echoing mountains, where they seem to be welcomed home. Then another and another peal, or rather crashing, splintering stroke, follows in quick succession, perchance splitting some giant pine or fir from top to bottom into long rails and slivers, and scattering them to all points of the compass. Now comes the rain, with corresponding extravagant grandeur, covering the ground high and low with a sheet of flowing water, a transparent film fitted like a skin upon the rugged anatomy of the landscape, making the rocks glitter and glow, gathering in the ravines, flooding the streams, and making them shout and boom in reply to the thunder.

How interesting to trace the history of a single raindrop! It is not long, geologically speaking, as we have seen, since the first raindrops fell on the newborn leafless Sierra landscapes. How different the lot of these falling now! Happy the showers that fall on so fair a wilderness, — scarce a single drop can fail to find a beautiful spot, — on the tops of the peaks, on the shining glacier pavements, on the great smooth domes, on forests and gardens and brushy moraines, plashing, glinting, pattering, laving. Some go to the high snowy fountains to swell their well-saved stores; some into the lakes, washing the mountain windows, patting their smooth glassy levels, making dimples and bubbles and spray; some into the waterfalls and cascades, as if eager to join in their dance and song and beat their foam yet finer; good luck and good work for the happy mountain raindrops, each one of them a high waterfall in itself, descending from the cliffs and hollows of the clouds to the cliffs and hollows of the rocks, out of the sky-thunder into the thunder of the falling rivers. Some, falling on meadows and bogs, creep silently out of sight to the grass roots, hiding softly as in a nest, slipping, oozing hither, thither, seeking and finding their appointed work. Some, descending through the spires of the woods, sift spray through the shining needles, whispering peace and good cheer to each one of them. Some drops with happy aim glint on the sides of crystals, — quartz, hornblende, garnet, zir-

con, tourmaline, feldspar, — patter on grains
of gold and heavy way-worn nuggets; some,
with blunt plap-plap and low bass drumming,
fall on the broad leaves of veratrum, saxifrage,
cypripedium. Some happy drops fall straight
into the cups of flowers, kissing the lips of lilies.
How far they have to go, how many cups to
fill, great and small, cells too small to be seen,
cups holding half a drop as well as lake basins
between the hills, each replenished with equal
care, every drop in all the blessed throng a silvery newborn star with lake and river, garden
and grove, valley and mountain, all that the
landscape holds reflected in its crystal depths,
God's messenger, angel of love sent on its way
with majesty and pomp and display of power
that make man's greatest shows ridiculous.

Now the storm is over, the sky is clear, the
last rolling thunder-wave is spent on the peaks,
and where are the raindrops now — what has
become of all the shining throng? In winged
vapor rising some are already hastening back
to the sky, some have gone into the plants,
creeping through invisible doors into the
round rooms of cells, some are locked in crystals of ice, some in rock crystals, some in
porous moraines to keep their small springs
flowing, some have gone journeying on in the
rivers to join the larger raindrop of the ocean.

From form to form, beauty to beauty, ever changing, never resting, all are speeding on with love's enthusiasm, singing with the stars the eternal song of creation.

July 20. Fine calm morning; air tense and clear; not the slightest breeze astir; everything shining, the rocks with wet crystals, the plants with dew, each receiving its portion of irised dewdrops and sunshine like living creatures getting their breakfast, their dew manna coming down from the starry sky like swarms of smaller stars. How wondrous fine are the particles in showers of dew, thousands required for a single drop, growing in the dark as silently as the grass! What pains are taken to keep this wilderness in health, — showers of snow, showers of rain, showers of dew, floods of light, floods of invisible vapor, clouds, winds, all sorts of weather, interaction of plant on plant, animal on animal, etc., beyond thought! How fine Nature's methods! How deeply with beauty is beauty overlaid! the ground covered with crystals, the crystals with mosses and lichens and low-spreading grasses and flowers, these with larger plants leaf over leaf with ever-changing color and form, the broad palms of the firs outspread over these, the azure dome over all like a bell-flower, and star above star.

128

THE YOSEMITE

Yonder stands the South Dome, its crown high above our camp, though its base is four thousand feet below us; a most noble rock, it seems full of thought, clothed with living light, no sense of dead stone about it, all spiritualized, neither heavy looking nor light, steadfast in serene strength like a god.

Our shepherd is a queer character and hard to place in this wilderness. His bed is a hollow made in red dry-rot punky dust beside a log which forms a portion of the south wall of the corral. Here he lies with his wonderful everlasting clothing on, wrapped in a red blanket, breathing not only the dust of the decayed wood but also that of the corral, as if determined to take ammoniacal snuff all night after chewing tobacco all day. Following the sheep he carries a heavy six-shooter swung from his belt on one side and his luncheon on the other. The ancient cloth in which the meat, fresh from the frying-pan, is tied serves as a filter through which the clear fat and gravy juices drip down on his right hip and leg in clustering stalactites. This oleaginous formation is soon broken up, however, and diffused and rubbed evenly into his scanty apparel, by sitting down, rolling over, crossing his legs while resting on logs, etc., making shirt and trousers water-tight and shiny. His trousers, in parti-

cular, have become so adhesive with the mixed
fat and resin that pine needles, thin flakes and
fibres of bark, hair, mica scales and minute
grains of quartz, hornblende, etc., feathers,
seed wings, moth and butterfly wings, legs and
antennæ of innumerable insects, or even whole
insects such as the small beetles, moths and
mosquitoes, with flower petals, pollen dust
and indeed bits of all plants, animals, and minerals of the region adhere to them and are
safely imbedded, so that though far from being a naturalist he collects fragmentary specimens of everything and becomes richer than
he knows. His specimens are kept passably
fresh, too, by the purity of the air and the
resiny bituminous beds into which they are
pressed. Man is a microcosm, at least our
shepherd is, or rather his trousers. These
precious overalls are never taken off, and nobody knows how old they are, though one may
guess by their thickness and concentric structure. Instead of wearing thin they wear thick,
and in their stratification have no small geological significance.

Besides herding the sheep, Billy is the
butcher, while I have agreed to wash the few
iron and tin utensils and make the bread.
Then, these small duties done, by the time the
sun is fairly above the mountain-tops I am

beyond the flock, free to rove and revel in the wilderness all the big immortal days.

Sketching on the North Dome. It commands views of nearly all the valley besides a few of the high mountains. I would fain draw everything in sight — rock, tree, and leaf. But little can I do beyond mere outlines, — marks with meanings like words, readable only to myself, — yet I sharpen my pencils and work on as if others might possibly be benefited. Whether these picture-sheets are to vanish like fallen leaves or go to friends like letters, matters not much; for little can they tell to those who have not themselves seen similar wildness, and like a language have learned it. No pain here, no dull empty hours, no fear of the past, no fear of the future. These blessed mountains are so compactly filled with God's beauty, no petty personal hope or experience has room to be. Drinking this champagne water is pure pleasure, so is breathing the living air, and every movement of limbs is pleasure, while the whole body seems to feel beauty when exposed to it as it feels the camp-fire or sunshine, entering not by the eyes alone, but equally through all one's flesh like radiant heat, making a passionate ecstatic pleasure-glow not explainable. One's body then seems homogeneous throughout, sound as a crystal.

MY FIRST SUMMER IN THE SIERRA

Perched like a fly on this Yosemite dome, I gaze and sketch and bask, oftentimes settling down into dumb admiration without definite hope of ever learning much, yet with the longing, unresting effort that lies at the door of hope, humbly prostrate before the vast display of God's power, and eager to offer self-denial and renunciation with eternal toil to learn any lesson in the divine manuscript.

It is easier to feel than to realize, or in any way explain, Yosemite grandeur. The magnitudes of the rocks and trees and streams are so delicately harmonized they are mostly hidden. Sheer precipices three thousand feet high are fringed with tall trees growing close like grass on the brow of a lowland hill, and extending along the feet of these precipices a ribbon of meadow a mile wide and seven or eight long, that seems like a strip a farmer might mow in less than a day. Waterfalls, five hundred to one or two thousand feet high, are so subordinated to the mighty cliffs over which they pour that they seem like wisps of smoke, gentle as floating clouds, though their voices fill the valley and make the rocks tremble. The mountains, too, along the eastern sky, and the domes in front of them, and the succession of smooth rounded waves between, swelling higher, higher, with dark woods in

132

their hollows, serene in massive exuberant bulk and beauty, tend yet more to hide the grandeur of the Yosemite temple and make it appear as a subdued subordinate feature of the vast harmonious landscape. Thus every attempt to appreciate any one feature is beaten down by the overwhelming influence of all the others. And, as if this were not enough, lo! in the sky arises another mountain range with topography as rugged and substantial-looking as the one beneath it — snowy peaks and domes and shadowy Yosemite valleys — another version of the snowy Sierra, a new creation heralded by a thunder-storm. How fiercely, devoutly wild is Nature in the midst of her beauty-loving tenderness! — painting lilies, watering them, caressing them with gentle hand, going from flower to flower like a gardener while building rock mountains and cloud mountains full of lightning and rain. Gladly we run for shelter beneath an overhanging cliff and examine the reassuring ferns and mosses, gentle love tokens growing in cracks and chinks. Daisies, too, and ivesias, confiding wild children of light, too small to fear. To these one's heart goes home, and the voices of the storm become gentle. Now the sun breaks forth and fragrant steam arises. The birds are out singing on the edges of the groves. The west is flaming in gold and purple, ready for the ceremony of the sunset, and back I go to camp with my notes and pictures, the best of them printed in my mind as dreams. A fruitful day, without measured beginning or ending. A terrestrial eternity. A gift of good God.

Wrote to my mother and a few friends, mountain hints to each. They seem as near as if within voice-reach or touch. The deeper the solitude the less the sense of loneliness, and the nearer our friends. Now bread and tea, fir bed and good-night to Carlo, a look at the sky lilies, and death sleep until the dawn of another Sierra to-morrow.

July 21. Sketching on the Dome — no rain; clouds at noon about quarter filled the sky, casting shadows with fine effect on the white mountains at the heads of the streams, and a soothing cover over the gardens during the warm hours.

Saw a common house-fly and a grasshopper and a brown bear. The fly and grasshopper paid me a merry visit on the top of the Dome, and I paid a visit to the bear in the middle of a small garden meadow between the Dome and the camp where he was standing alert among the flowers as if willing to be seen to advantage. I had not gone more than half a mile from camp this morning, when Carlo, who was trotting on a few yards ahead of me, came to a sudden, cautious standstill. Down went tail and ears, and forward went his knowing nose, while he seemed to be saying, "Ha, what's this? A bear, I guess." Then a cautious advance of a few steps, setting his feet down softly like a hunting cat, and questioning the air as to the scent he had caught until all doubt vanished. Then he came back to me, looked me in the face, and with his speaking eyes reported a bear near by; then led on softly, careful, like an experienced hunter, not to make the slightest noise, and frequently looking back as if whispering, "Yes, it's a bear; come and I'll show you." Presently we came to where the sunbeams were streaming through between the purple shafts of the firs, which showed that we were nearing an open spot, and here Carlo came behind me, evidently sure that the bear was very near. So I crept to a low ridge of moraine boulders on the edge of a narrow garden meadow, and in this meadow I felt pretty sure the bear must be. I was anxious to get a good look at the sturdy mountaineer without alarming him; so drawing myself up noiselessly back of one of the largest of the trees I peered past its bulging buttresses, exposing only a part of my head, and there stood neighbor Bruin within a stone's throw, his hips covered by tall grass and flowers, and his front feet on the trunk of a fir that had fallen out into the meadow, which raised his head so high that he seemed to be standing erect. He had not yet seen me, but was looking and listening attentively, showing that in some way he was aware of our approach. I watched his gestures and tried to make the most of my opportunity to learn what I could about him, fearing he would catch sight of me and run away. For I had been told that this sort of bear, the cinnamon, always ran from his bad brother man, never showing fight unless wounded or in defense of young. He made a telling picture standing alert in the sunny forest garden. How well he played his part, harmonizing in bulk and color and shaggy hair with the trunks of the trees and lush vegetation, as natural a feature as any other in the landscape. After examining at leisure, noting the sharp muzzle thrust inquiringly forward, the long shaggy hair on his broad chest, the stiff, erect ears nearly buried in hair, and the slow, heavy way he moved his head, I thought I should like to see his gait in running, so I made a sudden rush at him, shouting and swinging my hat to frighten him, expecting to see him make haste to get away. But to my dismay he did not run or show any sign of running. On the contrary, he stood his ground ready to fight and defend himself, lowered his head, thrust it forward, and looked sharply and fiercely at me. Then I suddenly began to fear that upon me would fall the work of running; but I was afraid to run, and therefore, like the bear, held my ground. We stood staring at each other in solemn silence within a dozen yards or thereabouts, while I fervently hoped that the power of the human eye over wild beasts would prove as great as it is said to be. How long our awfully strenuous interview lasted, I don't know; but at length in the slow fullness of time he pulled his huge paws down off the log, and with magnificent deliberation turned and walked leisurely up the meadow, stopping frequently to look back over his shoulder to see whether I was pursuing him, then moving on again, evidently neither fearing me very much nor trusting me. He was probably about five hundred pounds in weight, a broad, rusty bundle of ungovernable wildness, a happy fellow whose lines have fallen in pleasant places. The flowery glade in which I saw him so well, framed like a picture, is one of the best of all I have yet discovered, a conservatory of Nature's precious plant people.

137

Tall lilies were swinging their bells over that bear's back, with geraniums, larkspurs, columbines, and daisies brushing against his sides. A place for angels, one would say, instead of bears.

In the great cañons Bruin reigns supreme. Happy fellow, whom no famine can reach while one of his thousand kinds of food is spared him. His bread is sure at all seasons, ranged on the mountain shelves like stores in a pantry. From one to the other, up or down he climbs, tasting and enjoying each in turn in different climates, as if he had journeyed thousands of miles to other countries north or south to enjoy their varied productions. I should like to know my hairy brothers better — though after this particular Yosemite bear, my very neighbor, had sauntered out of sight this morning, I reluctantly went back to camp for the Don's rifle to shoot him, if necessary, in defense of the flock. Fortunately I could n't find him, and after tracking him a mile or two towards Mount Hoffman I bade him Godspeed and gladly returned to my work on the Yosemite Dome.

The house-fly also seemed at home and buzzed about me as I sat sketching, and enjoying my bear interview now it was over. I wonder what draws house-flies so far up the mountains, heavy gross feeders as they are, sensitive to cold, and fond of domestic ease. How have they been distributed from continent to continent, across seas and deserts and mountain chains, usually so influential in determining boundaries of species both of plants and animals. Beetles and butterflies are sometimes restricted to small areas. Each mountain in a range, and even the different zones of a mountain, may have its own peculiar species. But the house-fly seems to be everywhere. I wonder if any island in mid-ocean is flyless. The bluebottle is abundant in these Yosemite woods, ever ready with his marvelous store of eggs to make all dead flesh fly. Bumblebees are here, and are well fed on boundless stores of nectar and pollen. The honeybee, though abundant in the foothills, has not yet got so high. It is only a few years since the first swarm was brought to California.

A queer fellow and a jolly fellow is the grasshopper. Up the mountains he comes on excursions, how high I don't know, but at least as far and high as Yosemite tourists. I was much interested with the hearty enjoyment of the one that danced and sang for me on the Dome this afternoon. He seemed brimful of glad, hilarious energy, manifested by springing into the air to a height of twenty or thirty feet, then diving and springing up again and making a sharp musical rattle just as the lowest point in the descent was reached. Up and down a dozen times or so he danced and sang, then alighted to rest, then up and at it again. The curves he described in the air in diving and rattling resembled those made by cords hanging loosely and attached at the same height at the ends, the loops nearly covering each other. Braver, heartier, keener, care-free enjoyment of life I have never seen or heard in any creature, great or small. The life of this comic redlegs, the mountain's merriest child, seems to be made up of pure, condensed gayety. The Douglas squirrel is the only living creature that I can compare him with in exuberant, rollicking, irrepressible jollity. Wonderful that these sublime mountains are so loudly cheered and brightened by a creature so queer. Nature in him seems to be snapping her fingers in the face of all earthly dejection and melancholy with a boyish hip-hip-hurrah. How the sound is made I do not understand. When he was on the ground he made not the slightest noise, nor when he was simply flying from place to place, but only when diving in curves, the motion seeming to be required for the sound; for the more vigorous the diving the more ener-

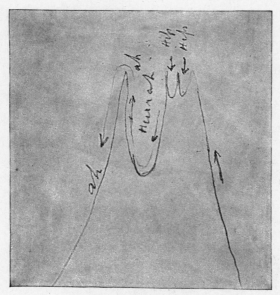

TRACK OF SINGING DANCING
GRASSHOPPER IN THE AIR
OVER NORTH DOME

getic the corresponding outbursts of jolly rattling. I tried to observe him closely while he was resting in the intervals of his performances; but he would not allow a near approach, always getting his jumping legs ready to spring for immediate flight, and keeping his eyes on me. A fine sermon the little fellow danced for me on the Dome, a likely place to look for sermons in stones, but not for grasshopper sermons. A large and imposing pulpit for so small a preacher. No danger of weakness in the knees of the world while Nature can spring such a rattle as this. Even the bear did not express for me the mountain's wild health and strength and happiness so tellingly as did this comical little hopper. No cloud of care in his day, no winter of discontent in sight. To him every day is a holiday; and when at length his sun sets, I fancy he will cuddle down on the forest floor and die like the leaves and flowers, and like them leave no unsightly remains calling for burial.

Sundown, and I must to camp. Good-night, friends three, — brown bear, rugged boulder of energy in groves and gardens fair as Eden; restless, fussy fly with gauzy wings stirring the air around all the world; and grasshopper, crisp, electric spark of joy enlivening the massy sublimity of the mountains like the laugh of a child. Thank you, thank you all three for your quickening company. Heaven guide every wing and leg. Good-night friends three, good-night.

July 22. A fine specimen of the black-tailed deer went bounding past camp this morning. A buck with wide spread of antlers, showing admirable vigor and grace. Wonderful the beauty, strength, and graceful movements of animals in wildernesses, cared for by Nature only, when our experience with domestic animals would lead us to fear that all the so-called neglected wild beasts would degenerate. Yet the upshot of Nature's method of breeding and teaching seems to lead to excellence of every sort. Deer, like all wild animals, are as clean as plants. The beauties of their gestures and attitudes, alert or in repose, surprise yet more than their bounding exuberant strength. Every movement and posture is graceful, the very poetry of manners and motion. Mother Nature is too often spoken of as in reality no mother at all. Yet how wisely, sternly, tenderly she loves and looks after her children in all sorts of weather and wildernesses. The more I see of deer the more I admire them as mountaineers. They make their way into the heart of the roughest solitudes with smooth reserve of strength, through dense belts of brush and for-

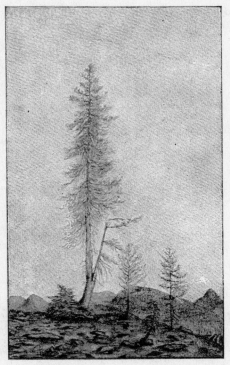

MT. CLARK   MT. STARR KING
TOP OF S. DOME
ABIES MAGNIFICA

est encumbered with fallen trees and boulder piles, across cañons, roaring streams, and snow-fields, ever showing forth beauty and courage. Over nearly all the continent the deer find homes. In the Florida savannas and hum-mocks, in the Canada woods, in the far north, roaming over mossy tundras, swimming lakes and rivers and arms of the sea from island to island washed with waves, or climbing rocky mountains, everywhere healthy and able, add-ing beauty to every landscape, — a truly admir-able creature and great credit to Nature.

Have been sketching a silver fir that stands on a granite ridge a few hundred yards to the eastward of camp — a fine tree with a particular snow-storm story to tell. It is about one hun-dred feet high, growing on bare rock, thrust-ing its roots into a weathered joint less than an inch wide, and bulging out to form a base to bear its weight. The storm came from the north while it was young and broke it down nearly to the ground, as is shown by the old, dead, weather-beaten top leaning out from the living trunk built up from a new shoot below the break. The annual rings of the trunk that have overgrown the dead sapling tell the year of the storm. Wonderful that a side branch forming a portion of one of the level collars that encircle the trunk of this species (*Abies magnifica*) should bend upward, grow erect, and take the place of the lost axis to form a new tree.

Many others, pines as well as firs, bear testimony to the crushing severity of this particular storm. Trees, some of them fifty to seventy-five feet high, were bent to the ground and buried like grass, whole groves vanishing as if the forest had been cleared away, leaving not a branch or needle visible until the spring thaw. Then the more elastic undamaged saplings rose again, aided by the wind, some reaching a nearly erect attitude, others remaining more or less bent, while those with broken backs endeavored to specialize a side branch below the break and make a leader of it to form a new axis of development. It is as if a man, whose back was broken or nearly so and who was compelled to go bent, should find a branch backbone sprouting straight up from below the break and should gradually develop new arms and shoulders and head, while the old damaged portion of his body died.

Grand white cloud mountains and domes created about noon as usual, ridges and ranges of endless variety, as if Nature dearly loved this sort of work, doing it again and again nearly every day with infinite industry, and producing beauty that never palls. A few zig-

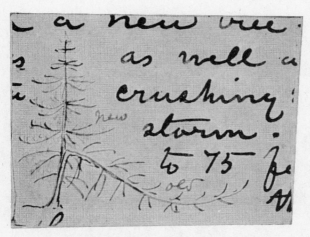

ILLUSTRATING GROWTH OF NEW PINE
FROM BRANCH BELOW THE BREAK OF
AXIS OF SNOW-CRUSHED TREE

zags of lightning, five minutes' shower, then a gradual wilting and clearing.

July 23. Another midday cloudland, displaying power and beauty that one never wearies in beholding, but hopelessly unsketchable and untellable. What can poor mortals say about clouds? While a description of their huge glowing domes and ridges, shadowy gulfs and cañons, and feather-edged ravines is being tried, they vanish, leaving no visible ruins. Nevertheless, these fleeting sky mountains are as substantial and significant as the more lasting upheavals of granite beneath them. Both alike are built up and die, and in God's calendar difference of duration is nothing. We can only dream about them in wondering, worshiping admiration, happier than we dare tell even to friends who see farthest in sympathy, glad to know that not a crystal or vapor particle of them, hard or soft, is lost; that they sink and vanish only to rise again and again in higher and higher beauty. As to our own work, duty, influence, etc., concerning which so much fussy pother is made, it will not fail of its due effect, though, like a lichen on a stone, we keep silent.

July 24. Clouds at noon occupying about half the sky gave half an hour of heavy rain to wash one of the cleanest landscapes in the

world. How well it is washed! The sea is hardly
less dusty than the ice-burnished pavements
and ridges, domes and cañons, and summit
peaks plashed with snow like waves with foam.
How fresh the woods are and calm after the
last films of clouds have been wiped from the
sky! A few minutes ago every tree was excited,
bowing to the roaring storm, waving, swirling,
tossing their branches in glorious enthusiasm
like worship. But though to the outer ear these
trees are now silent, their songs never cease.
Every hidden cell is throbbing with music
and life, every fibre thrilling like harp strings,
while incense is ever flowing from the balsam
bells and leaves. No wonder the hills and
groves were God's first temples, and the more
they are cut down and hewn into cathedrals
and churches, the farther off and dimmer seems
the Lord himself. The same may be said of
stone temples. Yonder, to the eastward of our
camp grove, stands one of Nature's cathedrals,
hewn from the living rock, almost conventional
in form, about two thousand feet high, nobly
adorned with spires and pinnacles, thrilling
under floods of sunshine as if alive like a grove-
temple, and well named "Cathedral Peak."
Even Shepherd Billy turns at times to this
wonderful mountain building, though apparently deaf to all stone sermons. Snow that re-

fused to melt in fire would hardly be more wonderful than unchanging dullness in the rays of God's beauty. I have been trying to get him to walk to the brink of Yosemite for a view, offering to watch the sheep for a day, while he should enjoy what tourists come from all over the world to see. But though within a mile of the famous valley, he will not go to it even out of mere curiosity. "What," says he, "is Yosemite but a cañon — a lot of rocks — a hole in the ground — a place dangerous about falling into — a d—d good place to keep away from." "But think of the waterfalls, Billy — just think of that big stream we crossed the other day, falling half a mile through the air — think of that, and the sound it makes. You can hear it now like the roar of the sea." Thus I pressed Yosemite upon him like a missionary offering the gospel, but he would have none of it. "I should be afraid to look over so high a wall," he said. "It would make my head swim. There is nothing worth seeing anywhere, only rocks, and I see plenty of them here. Tourists that spend their money to see rocks and falls are fools, that's all. You can't humbug me. I've been in this country too long for that." Such souls, I suppose, are asleep, or smothered and befogged beneath mean pleasures and cares.

July 25. Another cloudland. Some clouds

have an over-ripe decaying look, watery and bedraggled and drawn out into wind-torn shreds and patches, giving the sky a littered appearance; not so these Sierra summer mid-day clouds. All are beautiful with smooth definite outlines and curves like those of glacier-polished domes. They begin to grow about eleven o'clock, and seem so wonderfully near and clear from this high camp one is tempted to try to climb them and trace the streams that pour like cataracts from their shadowy fountains. The rain to which they give birth is often very heavy, a sort of waterfall as imposing as if pouring from rock mountains. Never in all my travels have I found anything more truly novel and interesting than these midday mountains of the sky, their fine tones of color, majestic visible growth, and ever-changing scenery and general effects, though mostly as well let alone as far as description goes. I oftentimes think of Shelley's cloud poem, "I sift the snow on the mountains below."

CHAPTER VI

MOUNT HOFFMAN AND LAKE TENAYA

July 26. Ramble to the summit of Mount
Hoffman, eleven thousand feet high, the highest
point in life's journey my feet have yet touched.
And what glorious landscapes are about me,
new plants, new animals, new crystals, and
multitudes of new mountains far higher than
Hoffman, towering in glorious array along
the axis of the range, serene, majestic, snow-
laden, sun-drenched, vast domes and ridges
shining below them, forests, lakes, and meadows in the hollows, the pure blue bell-flower
sky brooding them all, — a glory day of admission into a new realm of wonders as if Nature
had wooingly whispered, "Come higher."
What questions I asked, and how little I know
of all the vast show, and how eagerly, tremu-
lously hopeful of some day knowing more,
learning the meaning of these divine symbols
crowded together on this wondrous page.

Mount Hoffman is the highest part of a ridge
or spur about fourteen miles from the axis of
the main range, perhaps a remnant brought
into relief and isolated by unequal denudation.

The southern slopes shed their waters into Yosemite Valley by Tenaya and Dome Creeks, the northern in part into the Tuolumne River, but mostly into the Merced by Yosemite Creek. The rock is mostly granite, with some small piles and crests rising here and there in picturesque pillared and castellated remnants of red metamorphic slates. Both the granite and slates are divided by joints, making them separable into blocks like the stones of artificial masonry, suggesting the Scripture "He hath builded the mountains." Great banks of snow and ice are piled in hollows on the cool precipitous north side forming the highest perennial sources of Yosemite Creek. The southern slopes are much more gradual and accessible. Narrow slot-like gorges extend across the summit at right angles, which look like lanes, formed evidently by the erosion of less resisting beds. They are usually called "devil's slides," though they lie far above the region usually haunted by the devil; for though we read that he once climbed an exceeding high mountain, he cannot be much of a mountaineer, for his tracks are seldom seen above the timber-line.

The broad gray summit is barren and desolate-looking in general views, wasted by ages of gnawing storms; but looking at the surface in detail, one finds it covered by thousands

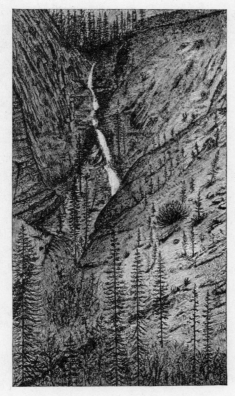

APPROACH OF DOME CREEK
TO YOSEMITE

and millions of charming plants with leaves
and flowers so small they form no mass of color
visible at a distance of a few hundred yards.
Beds of azure daisies smile confidingly in moist
hollows, and along the banks of small rills,
with several species of eriogonum, silky-leaved
ivesia, pentstemon, orthocarpus, and patches
of *Primula suffruticosa*, a beautiful shrubby
species. Here also I found bryanthus, a charming heathwort covered with purple flowers and
dark green foliage like heather, and three trees
new to me — a hemlock and two pines. The
hemlock (*Tsuga Mertensiana*) is the most
beautiful conifer I have ever seen; the branches
and also the main axis droop in a singularly
graceful way, and the dense foliage covers
the delicate, sensitive, swaying branchlets all
around. It is now in full bloom, and the
flowers, together with thousands of last season's cones still clinging to the drooping sprays,
display wonderful wealth of color, brown and
purple and blue. Gladly I climbed the first
tree I found to revel in the midst of it. How
the touch of the flowers makes one's flesh tingle! The pistillate are dark, rich purple, and
almost translucent, the staminate blue, — a
vivid, pure tone of blue like the mountain sky,
— the most uncommonly beautiful of all the
Sierra tree flowers I have seen. How wonder-

ful that, with all its delicate feminine grace and beauty of form and dress and behavior, this lovely tree up here, exposed to the wildest blasts, has already endured the storms of centuries of winters!

The two pines also are brave storm-enduring trees, the mountain pine (*Pinus monticola*) and the dwarf pine (*Pinus albicaulis*). The mountain pine is closely related to the sugar pine, though the cones are only about four to six inches long. The largest trees are from five to six feet in diameter at four feet above the ground, the bark rich brown. Only a few storm-beaten adventurers approach the summit of the mountain. The dwarf or whitebark pine is the species that forms the timberline, where it is so completely dwarfed that one may walk over the top of a bed of it as over snow-pressed chaparral.

How boundless the day seems as we revel in these storm-beaten sky gardens amid so vast a congregation of onlooking mountains! Strange and admirable it is that the more savage and chilly and storm-chafed the mountains, the finer the glow on their faces and the finer the plants they bear. The myriads of flowers tingeing the mountain-top do not seem to have grown out of the dry, rough gravel of disintegration, but rather they appear as visitors, a cloud of witnesses to Nature's love in what we in our timid ignorance and unbelief call howling desert. The surface of the ground, so dull and forbidding at first sight, besides being rich in plants, shines and sparkles with crystals: mica, hornblende, feldspar, quartz, tourmaline. The radiance in some places is so great as to be fairly dazzling, keen lance rays of every color flashing, sparkling in glorious abundance, joining the plants in their fine, brave beauty-work — every crystal, every flower a window opening into heaven, a mirror reflecting the Creator.

From garden to garden, ridge to ridge, I drifted enchanted, now on my knees gazing into the face of a daisy, now climbing again and again among the purple and azure flowers of the hemlocks, now down into the treasuries of the snow, or gazing afar over domes and peaks, lakes and woods, and the billowy glaciated fields of the upper Tuolumne, and trying to sketch them. In the midst of such beauty, pierced with its rays, one's body is all one tingling palate. Who would n't be a mountaineer! Up here all the world's prizes seem nothing.

The largest of the many glacier lakes in sight, and the one with the finest shore scenery, is Tenaya, about a mile long, with an imposing mountain dipping its feet into it on the south side, Cathedral Peak a few miles above its head, many smooth swelling rock-waves and domes on the north, and in the distance southward a multitude of snowy peaks, the fountain-heads of rivers. Lake Hoffman lies shimmering beneath my feet, mountain pines around its shining rim. To the north-ward the picturesque basin of Yosemite Creek glitters with lakelets and pools; but the eye is soon drawn away from these bright mirror wells, however attractive, to revel in the glorious congregation of peaks on the axis of the range in their robes of snow and light.

Carlo caught an unfortunate woodchuck when it was running from a grassy spot to its boulder-pile home — one of the hardiest of the mountain animals. I tried hard to save him, but in vain. After telling Carlo that he must be careful not to kill anything, I caught sight, for the first time, of the curious pika, or little chief hare, that cuts large quantities of lupines and other plants and lays them out to dry in the sun for hay, which it stores in underground barns to last through the long, snowy winter. Coming upon these plants freshly cut and lying in handfuls here and there on the rocks has a startling effect of busy life on the lonely mountain-top. These little haymakers,

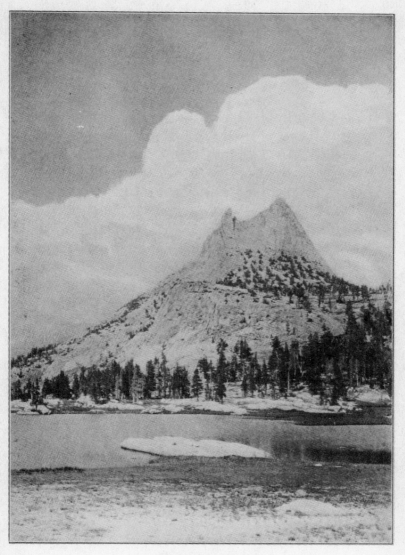

CATHEDRAL PEAK

endowed with brain stuff something like our own, — God up here looking after them, — what lessons they teach, how they widen our sympathy!

An eagle soaring above a sheer cliff, where I suppose its nest is, makes another striking show of life, and helps to bring to mind the other people of the so-called solitude — deer in the forest caring for their young; the strong, well-clad, well-fed bears; the lively throng of squirrels; the blessed birds, great and small, stirring and sweetening the groves; and the clouds of happy insects filling the sky with joyous hum as part and parcel of the down-pouring sunshine. All these come to mind, as well as the plant people, and the glad streams singing their way to the sea. But most impressive of all is the vast glowing countenance of the wilderness in awful, infinite repose.

Toward sunset, enjoyed a fine run to camp, down the long south slopes, across ridges and ravines, gardens and avalanche gaps, through the firs and chaparral, enjoying wild excitement and excess of strength, and so ends a day that will never end.

July 27. Up and away to Lake Tenaya, — another big day, enough for a lifetime. The rocks, the air, everything speaking with audible voice or silent; joyful, wonderful, enchanting, banishing weariness and sense of time. No longing for anything now or hereafter as we go home into the mountain's heart. The level sunbeams are touching the fir-tops, every leaf shining with dew. Am holding an easterly course, the deep cañon of Tenaya Creek on the right hand, Mount Hoffman on the left, and the lake straight ahead about ten miles distant, the summit of Mount Hoffman about three thousand feet above me, Tenaya Creek four thousand feet below and separated from the shallow, irregular valley, along which most of the way lies, by smooth domes and wave-ridges. Many mossy emerald bogs, meadows, and gardens in rocky hollows to wade and saunter through — and what fine plants they give me, what joyful streams I have to cross, and how many views are displayed of the Hoffman and Cathderal Peak masonry, and what a wondrous breadth of shining granite pavement to walk over for the first time about the shores of the lake! On I sauntered in freedom complete; body without weight as far as I was aware; now wading through starry parnassia bogs, now through gardens shoulder deep in larkspur and lilies, grasses and rushes, shaking off showers of dew; crossing piles of crystalline moraine boulders, bright mirror pavements, and cool, cheery streams going to

Yosemite; crossing bryanthus carpets and the scoured pathways of avalanches, and thickets of snow-pressed ceanothus; then down a broad, majestic stairway into the ice-sculptured lake-basin.

The snow on the high mountains is melting fast, and the streams are singing bank-full, swaying softly through the level meadows and bogs, quivering with sun-spangles, swirling in pot-holes, resting in deep pools, leaping, shouting in wild, exulting energy over rough boulder dams, joyful, beautiful in all their forms. No Sierra landscape that I have seen holds anything truly dead or dull, or any trace of what in manufactories is called rubbish or waste; everything is perfectly clean and pure and full of divine lessons. This quick, inevitable interest attaching to everything seems marvelous until the hand of God becomes visible; then it seems reasonable that what interests Him may well interest us. When we try to pick out anything by itself, we find it hitched to everything else in the universe. One fancies a heart like our own must be beating in every crystal and cell, and we feel like stopping to speak to the plants and animals as friendly fellow mountaineers. Nature as a poet, an enthusiastic workingman, becomes more and more visible the farther and higher we go; for the mountains are fountains — beginning places, however related to sources beyond mortal ken.

I found three kinds of meadows: (1) Those contained in basins not yet filled with earth enough to make a dry surface. They are planted with several species of carex, and have their margins diversified with robust flowering plants such as veratrum, larkspur, lupine, etc. (2) Those contained in the same sort of basins, once lakes like the first, but so situated in relation to the streams that flow through them and beds of transportable sand, gravel, etc., that they are now high and dry and well drained. This dry condition and corresponding difference in their vegetation may be caused by no superiority of position, or power of transporting filling material in the streams that belong to them, but simply by the basin being shallow and therefore sooner filled. They are planted with grasses, mostly fine, silky, and rather short-leaved, *Calamagrostis* and *Agrostis* being the principal genera. They form delightfully smooth, level sods in which one finds two or three species of gentian and as many of purple and yellow orthocarpus, violet, vaccinium, kalmia, bryanthus, and lonicera. (3) Meadows hanging on ridge and mountain slopes, not in basins at all, but made and held

in place by masses of boulders and fallen trees,
which, forming dams one above another in
close succession on small, outspread, chan-
nelless streams, have collected soil enough
for the growth of grasses, carices, and many
flowering plants, and being kept well watered,
without being subject to currents sufficiently
strong to carry them away, a hanging or sloping meadow is the result. Their surfaces are
seldom so smooth as the others, being roughened more or less by the projecting tops of the
dam rocks or logs; but at a little distance this
roughness is not noticed, and the effect is very
striking — bright green, fluent, down-sweeping flowery ribbons on gray slopes. The broad
shallow streams these meadows belong to are
mostly derived from banks of snow and because the soil is well drained in some places,
while in others the dam rocks are packed close
and caulked with bits of wood and leaves, making boggy patches; the vegetation, of course,
is correspondingly varied. I saw patches of
willow, bryanthus, and a fine show of lilies
on some of them, not forming a margin, but
scattered about among the carex and grass.
Most of these meadows are now in their prime.
How wonderful must be the temper of the
elastic leaves of grasses and sedges to make
curves so perfect and fine. Tempered a little

harder, they would stand erect, stiff and bristly, like strips of metal; a little softer, and every leaf would lie flat. And what fine painting and tinting there is on the glumes and pales, stamens and feathery pistils. Butterflies colored like the flowers waver above them in wonderful profusion, and many other beautiful winged people, numbered and known and loved only by the Lord, are waltzing together high over head, seemingly in pure play and hilarious enjoyment of their little sparks of life. How wonderful they are! How do they get a living, and endure the weather? How are their little bodies, with muscles, nerves, organs, kept warm and jolly in such admirable exuberant health? Regarded only as mechanical inventions, how wonderful they are! Compared with these, Godlike man's greatest machines are as nothing.

Most of the sandy gardens on moraines are in prime beauty like the meadows, though some on the north sides of rocks and beneath groves of sapling pines have not yet bloomed. On sunny sheets of crystal soil along the slopes of the Hoffman Mountains, I saw extensive patches of ivesia and purple gilia with scarce a green leaf, making fine clouds of color. Ribes bushes, vaccinium, and kalmia, now in flower, make beautiful rugs and borders along the banks of the streams. Shaggy beds of dwarf oak (*Quercus chrysolepis*, var. *vaccinifolia*) over which one may walk are common on rocky moraines, yet this is the same species as the large live oak seen near Brown's Flat. The most beautiful of the shrubs is the purple-flowered bryanthus, here making glorious carpets at an elevation of nine thousand feet.

The principal tree for the first mile or two from camp is the magnificent silver fir, which reaches perfection here both in size and form of individual trees, and in the mode of grouping in groves with open spaces between. So trim and tasteful are these silvery, spiry groves one would fancy they must have been placed in position by some master landscape gardener, their regularity seeming almost conventional. But Nature is the only gardener able to do work so fine. A few noble specimens two hundred feet high occupy central positions in the groups with younger trees around them; and outside of these another circle of yet smaller ones, the whole arranged like tastefully symmetrical bouquets, every tree fitting nicely the place assigned to it as if made especially for it; small roses and eriogonums are usually found blooming on the open spaces about the groves, forming charming pleasure grounds. Higher, the firs gradually become smaller and less perfect, many showing double summits, indicating storm stress. Still, where good moraine soil is found, even on the rim of the lake-basin, specimens one hundred and fifty feet in height and five feet in diameter occur nearly nine thousand feet above the sea. The saplings, I find, are mostly bent with the crushing weight of the winter snow, which at this elevation must be at least eight or ten feet deep, judging by marks on the trees; and this depth of compacted snow is heavy enough to bend and bury young trees twenty or thirty feet in height and hold them down for four or five months. Some are broken; the others spring up when the snow melts and at length attain a size that enables them to withstand the snow pressure. Yet even in trees five feet thick the traces of this early discipline are still plainly to be seen in their curved insteps, and frequently in old dried saplings protruding from the trunk, partially overgrown by the new axis developed from a branch below the break. Yet through all this stress the forest is maintained in marvelous beauty.

Beyond the silver firs I find the two-leaved pine (*Pinus contorta*, var. *Murrayana*) forms the bulk of the forest up to an elevation of ten thousand feet or more — the highest timber-belt of the Sierra. I saw a specimen nearly five feet in diameter growing on deep, well-watered soil at an elevation of about nine thousand feet. The form of this species varies very much with position, exposure, soil, etc. On stream-banks, where it is closely planted, it is very slender; some specimens seventy-five feet high do not exceed five inches in diameter at the ground, but the ordinary form, as far as I have seen, is well proportioned. The average diameter when full grown at this elevation is about twelve or fourteen inches, height forty or fifty feet, the straggling branches bent up at the end, the bark thin and bedraggled with amber-colored resin. The pistillate flowers form little crimson rosettes a fourth of an inch in diameter on the ends of the branchlets, mostly hidden in the leaf-tassels; the staminate are about three eighths of an inch in diameter, sulphur-yellow, in showy clusters, giving a remarkably rich effect — a brave, hardy mountaineer pine, growing cheerily on rough beds of avalanche boulders and joints of rock pavements, as well as in fertile hollows, standing up to the waist in snow every winter for centuries, facing a thousand storms and blooming every year in colors as bright as those worn by the sun-drenched trees of the tropics.

A still hardier mountaineer is the Sierra juniper (*Juniperus occidentalis*), growing mostly on domes and ridges and glacier pavements. A thickset, sturdy, picturesque highlander, seemingly content to live for more than a score of centuries on sunshine and snow; a truly wonderful fellow, dogged endurance expressed in every feature, lasting about as long as the granite he stands on. Some are nearly as broad as high. I saw one on the shore of the lake nearly ten feet in diameter, and many six to eight feet. The bark, cinnamon-colored, flakes off in long ribbon-like strips with a satiny luster. Surely the most enduring of all tree mountaineers, it never seems to die a natural death, or even to fall after it has been killed. If protected from accidents, it would perhaps be immortal. I saw some that had withstood an avalanche from snowy Mount Hoffman cheerily putting out new branches, as if repeating, like Grip, "Never say die." Some were simply standing on the pavement where no fissure more than half an inch wide offered a hold for its roots. The common height for these rock-dwellers is from ten to twenty feet; most of the old ones have broken tops, and are mere stumps, with a few tufted branches, forming picturesque brown pillars on bare pavements, with plenty of elbow-room and a clear view in every direction. On good moraine soil it reaches a height of from forty to

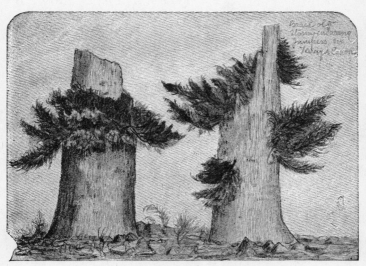

JUNIPERS IN TENAYA CAÑON

sixty feet, with dense gray foliage. The rings of the trunk are very thin, eighty to an inch of diameter in some specimens I examined. Those ten feet in diameter must be very old — thousands of years. Wish I could live, like these junipers, on sunshine and snow, and stand beside them on the shore of Lake Tenaya for a thousand years. How much I should see, and how delightful it would be! Everything in the mountains would find me and come to me, and everything from the heavens like light.

The lake was named for one of the chiefs of the Yosemite tribe. Old Tenaya is said to have been a good Indian to his tribe. When a company of soldiers followed his band into Yosemite to punish them for cattle-stealing and other crimes, they fled to this lake by a trail that leads out of the upper end of the valley, early in the spring, while the snow was still deep; but being pursued, they lost heart and surrendered. A fine monument the old man has in this bright lake, and likely to last a long time, though lakes die as well as Indians, being gradually filled with detritus carried in by the feeding streams, and to some extent also by snow avalanches and rain and wind. A considerable portion of the Tenaya basin is already changed into a forested flat and meadow at the upper end, where the main tributary enters from Cathedral Peak. Two other tributaries come from the Hoffman Range. The outlet flows westward through Tenaya Cañon to join the Merced River in Yosemite. Scarce a handful of loose soil is to be seen on the north shore. All is bare, shining granite, suggesting the Indian name of the lake, Pywiack, meaning shining rock. The basin seems to have been slowly excavated by the ancient glaciers, a marvelous work requiring countless thousands of years. On the south side an imposing mountain rises from the water's edge to a height of three thousand feet or more, feathered with hemlock and pine; and huge shining domes on the east, over the tops of which the grinding, wasting, molding glacier must have swept as the wind does to-day.

July 28. No cloud mountains, only curly currus wisps scarce perceptible, and the want of thunder to strike the noon hour seems strange, as if the Sierra clock had stopped. Have been studying the *magnifica* fir — measured one near two hundred and forty feet high, the tallest I have yet seen. This species is the most symmetrical of all conifers, but though gigantic in size it seldom lives more than four or five hundred years. Most of the trees die from the attacks of a fungus at the age of two or three centuries. This dry-rot fungus perhaps enters the trunk by way of the stumps of limbs broken off by the snow that loads the broad palmate branches. The younger specimens are marvels of symmetry, straight and erect as a plumb-line, their branches in regular level whorls of five mostly, each branch as exact in its divisions as a fern frond, and thickly covered by the leaves, making a rich plush over all the tree, excepting only the trunk and a small portion of the main limbs. The leaves turn upward, especially on the branchlets, and are stiff and sharp, pointed on all the upper portion of the tree. They remain on the tree about eight or ten years, and as the growth is rapid it is not rare to find the leaves still in place on the upper part of the axis where it is three to four inches in diameter, wide apart of course, and their spiral arrangement beautifully displayed. The leaf-scars are conspicuous for twenty years or more, but there is a good deal of variation in different trees as to the thickness and sharpness of the leaves.

After the excursion to Mount Hoffman I had seen a complete cross-section of the Sierra forest, and I find that *Abies magnifica* is the most symmetrical tree of all the noble coniferous company. The cones are grand affairs, superb in form, size, and color, cylindrical, stand erect on the upper branches like casks, and are from five to eight inches in length by three or four in diameter, greenish gray, and covered with fine down which has a silvery luster in the sunshine, and their brilliance is augmented by beads of transparent balsam which seems to have been poured over each cone, bringing to mind the old ceremonies of anointing with oil. If possible, the inside of the cone is more beautiful than the outside; the scales, bracts, and seed wings are tinted with the loveliest rosy purple with a bright lustrous iridescence; the seeds, three fourths of an inch long, are dark brown. When the cones are ripe the scales and bracts fall off, setting the seeds free to fly to their predestined places, while the dead spike-like axes are left on the branches for many years to mark the positions of the vanished cones, excepting those cut off when green by the Douglas squirrel. How he gets his teeth under the broad bases of the sessile cones, I don't know. Climbing these trees on a sunny day to visit the growing cones and to gaze over the tops of the forest is one of my best enjoyments.

July 29. Bright, cool, exhilarating. Clouds about .05. Another glorious day of rambling, sketching, and universal enjoyment.

July 30. Clouds .20, but the regular shower did not reach us, though thunder was heard a few miles off striking the noon hour. Ants, flies, and mosquitoes seem to enjoy this fine climate. A few house-flies have discovered our camp. The Sierra mosquitoes are courageous and of good size, some of them measuring nearly an inch from tip of sting to tip of folded wings. Though less abundant than in most wildernesses, they occasionally make quite a hum and stir, and pay but little attention to time or place. They sting anywhere, any time of day, wherever they can find anything worth while, until they are themselves stung by frost. The large, jet-black ants are only ticklish and troublesome when one is lying down under the trees. Noticed a borer drilling a silver fir. Ovipositor about an inch and a half in length, polished and straight like a needle. When not in use, it is folded back in a sheath, which extends straight behind like the legs of a crane in flying. This drilling, I suppose, is to save nest building, and the after care of feeding the young. Who would guess that in the brain of a fly so much knowledge could find lodgment? How do they know that their eggs will hatch in such holes, or, after they hatch, that the soft, helpless grubs will find the right sort of nourishment in silver fir sap? This domestic arrangement calls to mind the curious family of gallflies. Each species seems to know what kind of plant will respond to the irritation or stimulus of the puncture it makes and the eggs it lays, in forming a growth that not only answers for a nest and home but also provides food for the young. Probably these gallflies make mistakes at times, like anybody else; but when they do, there is simply a failure of that particular brood, while enough to perpetuate the species do find the proper plants and nourishment. Many mistakes of this kind might be made without being discovered by us. Once a pair of wrens made the mistake of building a nest in the sleeve of a workman's coat, which was called for at sundown, much to the consternation and discomfiture of the birds. Still the marvel remains that any of the children of such small people as gnats and mosquitoes should escape their own and their parents' mistakes, as well as the vicissitudes of the weather and hosts of enemies, and come forth in full vigor and perfection to enjoy the sunny world. When we think of the small creatures that are visible, we are led to think of many that are smaller still and lead us on and on into infinite mystery.

July 31. Another glorious day, the air as delicious to the lungs as nectar to the tongue;

indeed the body seems one palate, and tingles equally throughout. Cloudiness about .05, but our ordinary shower has not yet reached us, though I hear thunder in the distance.

The cheery little chipmunk, so common about Brown's Flat, is common here also, and perhaps other species. In their light, airy habits they recall the familiar species of the Eastern States, which we admired in the oak openings of Wisconsin as they skimmed along the zigzag rail fences. These Sierra chipmunks are more arboreal and squirrel-like. I first noticed them on the lower edge of the coniferous belt, where the Sabine and yellow pines meet, — exceedingly interesting little fellows, full of odd, funny ways, and without being true squirrels, have most of their acomplishments without their aggressive quarrelsomeness. I never weary watching them as they frisk about in the bushes gathering seeds and berries, like song sparrows poising daintily on slender twigs, and making even less stir than most birds of the same size. Few of the Sierra animals interest me more; they are so able, gentle, confiding, and beautiful, they take one's heart, and get themselves adopted as darlings. Though weighing hardly more than field mice, they are laborious collectors of seeds, nuts, and cones, and are therefore well fed, but never in the least swollen with fat or lazily full. On the contrary, of their frisky, birdlike liveliness there is no end. They have a great variety of notes corresponding with their movements, some sweet and liquid, like water dripping with tinkling sounds into pools. They seem dearly to love teasing a dog, coming frequently almost within reach, then frisking away with lively chipping, like sparrows, beating time to their music with their tails, which at each chip describe half circles from side to side. Not even the Douglas squirrel is surer-footed or more fearless. I have seen them running about on sheer precipices of the Yosemite walls seemingly holding on with as little effort as flies, and as unconscious of danger, where, if the slightest slip were made, they would have fallen two or three thousand feet. How fine it would be could we mountaineers climb these tremendous cliffs with the same sure grip! The venture I made the other day for a view of the Yosemite Fall, and which tried my nerves so sorely, this little Tamias would have made for an ear of grass.

The woodchuck (*Arctomys monax*) of the bleak mountain-tops is a very different sort of mountaineer — the most bovine of rodents, a heavy eater, fat, aldermanic in bulk and fairly bloated, in his high pastures, like a cow in a clover field. One woodchuck would outweigh a hundred chipmunks, and yet he is by no means a dull animal. In the midst of what we regard as storm-beaten desolation he pipes and whistles right cheerily, and enjoys long life in his skyland homes. His burrow is made in disintegrated rocks or beneath large boulders. Coming out of his den in the cold hoarfrost mornings, he takes a sun-bath on some favorite flat-topped rock, then goes to breakfast in garden hollows, eats grass and flowers until comfortably swollen, then goes a-visiting to fight and play. How long a woodchuck lives in this bracing air I don't know, but some of them are rusty and gray like lichen-covered boulders.

August 1. A grand cloudland and five-minute shower, refreshing the blessed wilderness, already so fragrant and fresh, steeping the black meadow mold and dead leaves like tea.

The waycup, or flicker, so familiar to every boy in the old Middle West States, is one of the most common of the wood-peckers hereabouts, and makes one feel at home. I can see no difference in plumage or habits from the Eastern species, though the climate here is so different, — a fine, brave, confiding, beautiful bird. The robin, too, is here, with all his familiar notes and gestures, tripping daintily on open garden spots and high meadows. Over all

America he seems to be at home, moving from the plains to the mountains and from north to south, back and forth, up and down, with the march of the seasons and food supply. How admirable the constitution and temper of this brave singer, keeping in cheery health over so vast and varied a range! Oftentimes, as I wander through these solemn woods, awe-stricken and silent, I hear the reassuring voice of this fellow wanderer ringing out, sweet and clear, "Fear not! fear not!"

The mountain quail (*Oreortyx ricta*) I often meet in my walks — a small brown partridge with a very long, slender, ornamental crest worn jauntily like a feather in a boy's cap, giving it a very marked appearance. This species is considerably larger than the valley quail, so common on the hot foothills. They seldom alight in trees, but love to wander in flocks of from five or six to twenty through the ceanothus and manzanita thickets and over open, dry meadows and rocks of the ridges where the forest is less dense or wanting, uttering a low clucking sound to enable them to keep together. When disturbed they rise with a strong birr of wing-beats, and scatter as if exploded to a distance of a quarter of a mile or so. After the danger is past they call one another together with a loud piping note — Nature's beautiful mountain chickens. I have not yet found their nests. The young of this season are already hatched and away — new broods of happy wanderers half as large as their parents. I wonder how they live through the long winters, when the ground is snow-covered ten feet deep. They must go down towards the lower edge of the forest, like the deer, though I have not heard of them there.

The blue, or dusky, grouse is also common here. They like the deepest and closest fir woods, and when disturbed, burst from the branches of the trees with a strong, loud whir of wing-beats, and vanish in a wavering, silent slide, without moving a feather — a stout, beautiful bird about the size of the prairie chicken of the old west, spending most of the time in the trees, excepting the breeding season, when it keeps to the ground. The young are now able to fly. When scattered by man or dog, they keep still until the danger is supposed to be passed, then the mother calls them together. The chicks can hear the call a distance of several hundred yards, though it is not loud. Should the young be unable to fly, the mother feigns desperate lameness or death to draw one away, throwing herself at one's feet within two or three yards, rolling over on her back, kicking and gasping, so as to deceive man or beast. They are said to stay all the year in the woods hereabouts, taking shelter in dense tufted branches of fir and yellow pine during snowstorms, and feeding on the young buds of these trees. Their legs are feathered down to their toes, and I have never heard of their suffering in any sort of weather. Able to live on pine and fir buds, they are forever independent in the matter of food, which troubles so many of us and controls our movements. Gladly, if I could, I would live forever on pine buds, however full of turpentine and pitch, for the sake of this grand independence. Just to think of our sufferings last month merely for grist-mill flour. Man seems to have more difficulty in gaining food than any other of the Lord's creatures. For many in towns it is a consuming, lifelong struggle; for others, the danger of coming to want is so great, the deadly habit of endless hoarding for the future is formed, which smothers all real life, and is continued long after every reasonable need has been over-supplied.

On Mount Hoffman I saw a curious dove-colored bird that seemed half woodpecker, half magpie, or crow. It screams something like a crow, but flies like a woodpecker, and has a long, straight bill, with which I saw it opening the cones of the mountain and white-barked pines. It seems to keep to the heights, though no doubt it comes down for shelter during winter, if not for food. So far as food is concerned, these bird-mountaineers, I guess, can glean nuts enough, even in winter, from the different kinds of conifers; for always there are a few that have been unable to fly out of the cones and remain for hungry winter gleaners.

CHAPTER VII

A STRANGE EXPERIENCE

August 2. Clouds and showers, about the same as yesterday. Sketching all day on the North Dome until four or five o'clock in the afternoon, when, as I was busily employed thinking only of the glorious Yosemite landscape, trying to draw every tree and every line and feature of the rocks, I was suddenly, and without warning, possessed with the notion that my friend, Professor J. D. Butler, of the State University of Wisconsin, was below me in the valley, and I jumped up full of the idea of meeting him, with almost as much startling excitement as if he had suddenly touched me to make me look up. Leaving my work without the slightest deliberation, I ran down the western slope of the Dome and along the brink of the valley wall, looking for a way to the bottom, until I came to a side cañon, which, judging by its apparently continuous growth of trees and bushes, I thought might afford a practical way into the valley, and immediately began to make the descent, late as it was, as if drawn irresistibly. But after a little, com-

mon sense stopped me and explained that it
would be long after dark ere I could possibly
reach the hotel, that the visitors would be
asleep, that nobody would know me, that I
had no money in my pockets, and moreover
was without a coat. I therefore compelled
myself to stop, and finally succeeded in rea-
soning myself out of the notion of seeking my
friend in the dark, whose presence I only felt
in a strange, telepathic way. I succeeded in
dragging myself back through the woods to
camp, never for a moment wavering, however,
in my determination to go down to him next
morning. This I think is the most unexplainable notion that ever struck me. Had some
one whispered in my ear while I sat on the
Dome, where I had spent so many days, that
Professor Butler was in the valley, I could
not have been more surprised and startled.
When I was leaving the university, he said,
"Now, John, I want to hold you in sight and
watch your career. Promise to write me at
least once a year." I received a letter from
him in July, at our first camp in the Hollow,
written in May, in which he said that he might
possibly visit California some time this summer, and therefore hoped to meet me. But
inasmuch as he named no meeting-place, and
gave no directions as to the course he would

probably follow, and as I should be in the
wilderness all summer, I had not the slightest
hope of seeing him, and all thought of the
matter had vanished from my mind until this
afternoon, when he seemed to be wafted bodily
almost against my face. Well, to-morrow I
shall see; for, reasonable or unreasonable, I
feel I must go.

August 3. Had a wonderful day. Found
Professor Butler as the compass-needle finds
the pole. So last evening's telepathy, transcendental revelation, or whatever else it may
be called, was true; for, strange to say, he had
just entered the valley by way of the Coulterville Trail and was coming up the valley
past El Capitan when his presence struck me.
Had he then looked toward the North Dome
with a good glass when it first came in sight,
he might have seen me jump up from my work
and run toward him. This seems the one well-
defined marvel of my life of the kind called
supernatural; for, absorbed in glad Nature,
spirit-rappings, second sight, ghost stories,
etc., have never interested me since boyhood,
seeming comparatively useless and infinitely
less wonderful than Nature's open, harmonious, songful, sunny, everyday beauty.

This morning, when I thought of having
to appear among tourists at a hotel, I was

troubled because I had no suitable clothes, and at best am desperately bashful and shy. I was determined to go, however, to see my old friend after two years among strangers; got on a clean pair of overalls, a cashmere shirt, and a sort of jacket, — the best my camp wardrobe afforded, — tied my notebook on my belt, and strode away on my strange journey, followed by Carlo. I made my way though the gap discovered last evening, which proved to be Indian Cañon. There was no trail in it, and the rocks and brush were so rough that Carlo frequently called me back to help him down precipitous places. Emerging from the cañon shadows, I found a man making hay on one of the meadows, and asked him whether Professor Butler was in the valley. "I don't know," he replied; "but you can easily find out at the hotel. There are but few visitors in the valley just now. A small party came in yesterday afternoon, and I heard some one called Professor Butler, or Butterfield, or some name like that."

In front of the gloomy hotel I found a tourist party adjusting their fishing tackle. They all stared at me in silent wonderment, as if I had been seen dropping down through the trees from the clouds, mostly, I suppose, on account of my strange garb. Inquiring for the office, I was told it was locked, and that the landlord was away, but I might find the landlady, Mrs. Hutchings, in the parlor. I entered in a sad state of embarrassment, and after I had waited in the big, empty room and knocked at several doors the landlady at length appeared, and in reply to my question said she rather thought Professor Butler *was* in the valley, but to make sure, she would bring the register from the office. Among the names of the last arrivals I soon discovered the Professor's familiar handwriting, at the sight of which bashfulness vanished; and having learned that his party had gone up the valley, — probably to the Vernal and Nevada Falls, — I pushed on in glad pursuit, my heart now sure of its prey. In less than an hour I reached the head of the Nevada Cañon at the Vernal Fall, and just outside of the spray discovered a distinguished-looking gentleman, who, like everybody else I have seen to-day, regarded me curiously as I approached. When I made bold to inquire if he knew where Professor Butler was, he seemed yet more curious to know what could possibly have happened that required a messenger for the Professor, and instead of answering my question he asked with military sharpness, "Who wants him?" "I want him," I replied with equal sharp-

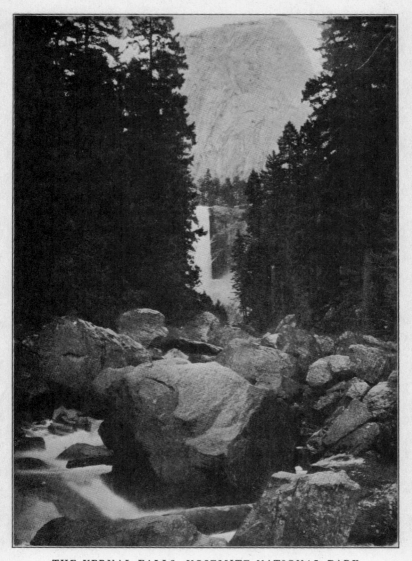

THE VERNAL FALLS, YOSEMITE NATIONAL PARK

ness. "Why? Do *you* know him?" "Yes,"
I said. "Do *you* know him?" Astonished that
any one in the mountains could possibly know
Professor Butler and find him as soon as he
had reached the valley, he came down to meet
the strange mountaineer on equal terms, and
courteously replied, "Yes, I know Professor
Butler very well. I am General Alvord, and
we were fellow students in Rutland, Vermont,
long ago, when we were both young." "But
where is he now?" I persisted, cutting short
his story. "He has gone beyond the falls with
a companion, to try to climb that big rock, the
top of which you see from here." His guide
now volunteered the information that it was
the Liberty Cap Professor Butler and his companion had gone to climb, and that if I waited
at the head of the fall I should be sure to find
them on their way down. I therefore climbed
the ladders alongside the Vernal Fall, and was
pushing forward, determined to go to the top
of Liberty Cap rock in my hurry, rather than
wait, if I should not meet my friend sooner.
So heart-hungry at times may one be to see a
friend in the flesh, however happily full and
care-free one's life may be. I had gone but
a short distance, however, above the brow of
the Vernal Fall when I caught sight of him in
the brush and rocks, half erect, groping his

way, his sleeves rolled up, vest open, hat in
his hand, evidently very hot and tired. When
he saw me coming he sat down on a boulder
to wipe the perspiration from his brow and
neck, and taking me for one of the valley
guides, he inquired the way to the fall ladders.
I pointed out the path marked with little piles
of stones, on seeing which he called his companion, saying that the way was found; but
he did not yet recognize me. Then I stood
directly in front of him, looked him in the face,
and held out my hand. He thought I was offering to assist him in rising. "Never mind," he
said. Then I said, "Professor Butler, don't
you know me?" "I think not," he replied;
but catching my eye, sudden recognition followed, and astonishment that I should have
found him just when he was lost in the brush
and did not know that I was within hundreds
of miles of him. "John Muir, John Muir,
where have you come from?" Then I told
him the story of my feeling his presence when
he entered the valley last evening, when he
was four or five miles distant, as I sat sketching on the North Dome. This, of course, only
made him wonder the more. Below the foot
of the Vernal Fall the guide was waiting with
his saddle-horse, and I walked along the trail,
chatting all the way back to the hotel, talking

of school days, friends in Madison, of the students, how each had prospered, etc., ever and anon gazing at the stupendous rocks about us, now growing indistinct in the gloaming, and again quoting from the poets — a rare ramble.

It was late ere we reached the hotel, and General Alvord was waiting the Professor's arrival for dinner. When I was introduced he seemed yet more astonished than the Professor at my descent from cloudland and going straight to my friend without knowing in any ordinary way that he was even in California. They had come on direct from the East, had not yet visited any of their friends in the state, and considered themselves undiscoverable. As we sat at dinner, the General leaned back in his chair, and looking down the table, thus introduced me to the dozen guests or so, including the staring fisherman mentioned above: "This man, you know, came down out of these huge, trackless mountains, you know, to find his friend Professor Butler here, the very day he arrived; and how did he know he was here? He just felt him, he says. This is the queerest case of Scotch farsightedness I ever heard of," etc., etc. While my friend quoted Shakespeare: "More things in heaven and earth, Horatio, than are dreamt of in your philosophy," "As the sun, ere he has risen, sometimes paints his image in the firmament, e'en so the shadows of events precede the events, and in to-day already walks to-morrow."

Had a long conversation, after dinner, over Madison days. The Professor wants me to promise to go with him, sometime, on a camping trip in the Hawaiian Islands, while I tried to get him to go back with me to camp in the high Sierra. But he says, "Not now." He must not leave the General; and I was surprised to learn they are to leave the valley tomorrow or next day. I'm glad I'm not great enough to be missed in the busy world.

August 4. It seemed strange to sleep in a paltry hotel chamber after the spacious magnificence and luxury of the starry sky and silver fir grove. Bade farewell to my friend and the General. The old soldier was very kind, and an interesting talker. He told me long stories of the Florida Seminole war, in which he took part, and invited me to visit him in Omaha. Calling Carlo, I scrambled home through the Indian Cañon gate, rejoicing, pitying the poor Professor and General, bound by clocks, almanacs, orders, duties, etc., and compelled to dwell with lowland care and dust and din, where Nature is covered and her voice smothered, while the poor, insignificant wanderer enjoys the freedom and glory of God's wilderness.

Apart from the human interest of my visit to-day, I greatly enjoyed Yosemite, which I had visited only once before, having spent eight days last spring in rambling amid its rocks and waters. Wherever we go in the mountains, or indeed in any of God's wild fields, we find more than we seek. Descending four thousand feet in a few hours, we enter a new world — climate, plants, sounds, inhabitants, and scenery all new or changed. Near camp the goldcup oak forms sheets of chaparral, on top of which we may make our beds. Going down the Indian Cañon we observe this little bush changing by regular gradations to a large bush, to a small tree, and then larger, until on the rocky taluses near the bottom of the valley we find it developed into a broad, wide-spreading, gnarled, picturesque tree from four to eight feet in diameter, and forty or fifty feet high. Innumerable are the forms of water displayed. Every gliding reach, cascade, and fall has characters of its own. Had a good view of the Vernal and Nevada, two of the main falls of the valley, less than a mile apart, and offering striking differences in voice, form, color, etc. The Vernal, four hundred feet high and about seventy-five or eighty feet wide, drops smoothly over a round-lipped precipice and forms a superb apron of embroidery, green and white, slightly folded and fluted, maintaining this form nearly to the bottom, where it is suddenly veiled in quick-flying billows of spray and mist, in which the afternoon sunbeams play with ravishing beauty of rainbow colors. The Nevada is white from its first appearance as it leaps out into the freedom of the air. At the head it presents a twisted appearance, by an overfolding of the current from striking on the side of its channel just before the first free out-bounding leap is made. About two thirds of the way down, the hurrying throng of comet-shaped masses glance on an inclined part of the face of the precipice and are beaten into yet whiter foam, greatly expanded, and sent bounding outward, making an indescribably glorious show, especially when the afternoon sunshine is pouring into it. In this fall — one of the most wonderful in the world — the water does not seem to be under the dominion of ordinary laws, but rather as if it were a living creature, full of the strength of the mountains and their huge, wild joy.

From beneath heavy throbbing blasts of spray the broken river is seen emerging in ragged boulder-chafed strips. These are speedily gathered into a roaring torrent, showing that the young river is still gloriously alive. On it goes, shouting, roaring, exulting in its strength, passes through a gorge with sublime display of energy, then suddenly expands on a gently inclined pavement, down which it rushes in thin sheets and folds of lace-work into a quiet pool, — "Emerald Pool," as it is called, — a stopping-place, a period separating two grand sentences. Resting here long enough to part with its foam-bells and gray mixtures of air, it glides quietly to the verge of the Vernal precipice in a broad sheet and makes its new display in the Vernal Fall; then more rapids and rock tossings down the cañon, shaded by live oak, Douglas spruce, fir, maple, and dogwood. It receives the Illilouette tributary, and makes a long sweep out into the level, sun-filled valley to join the other streams which, like itself, have danced and sung their way down from snowy heights to form the main Merced — the river of Mercy. But of this there is no end, and life, when one thinks of it, is so short. Never mind, one day in the midst of these divine glories is well worth living and toiling and starving for.

Before parting with Professor Butler he gave me a book, and I gave him one of my pencil sketches for his little son Henry, who

is a favorite of mine. He used to make many
visits to my room when I was a student. Never
shall I forget his patriotic speeches for the
Union, mounted on a tall stool, when he was
only six years old.

It seems strange that visitors to Yosemite
should be so little influenced by its novel grandeur, as if their eyes were bandaged and their
ears stopped. Most of those I saw yesterday
were looking down as if wholly unconscious of
anything going on about them, while the sublime rocks were trembling with the tones of the
mighty chanting congregation of waters gathered from all the mountains round about, making music that might draw angels out of heaven.
Yet respectable-looking, even wise-looking people were fixing bits of worms on bent pieces
of wire to catch trout. Sport they called it.
Should church-goers try to pass the time fishing in baptismal fonts while dull sermons were
being preached, the so-called sport might not
be so bad; but to play in the Yosemite temple,
seeking pleasure in the pain of fishes struggling
for their lives, while God himself is preaching
his sublimest water and stone sermons!

Now I'm back at the camp-fire, and cannot
help thinking about my recognition of my
friend's presence in the valley while he was four
or five miles away, and while I had no means of

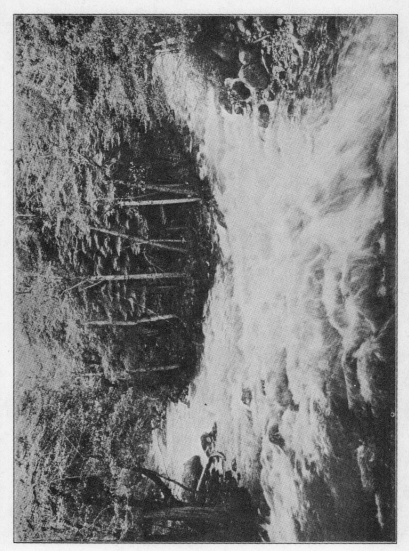

THE HAPPY ISLES, YOSEMITE NATIONAL PARK

knowing that he was not thousands of miles away. It seems supernatural, but only because it is not understood. Anyhow, it seems silly to make so much of it, while the natural and common is more truly marvelous and mysterious than the so-called supernatural. Indeed most of the miracles we hear of are infinitely less wonderful than the commonest of natural phenomena, when fairly seen. Perhaps the invisible rays that struck me while I sat at work on the Dome are something like those which attract and repel people at first sight, concerning which so much nonsense has been written. The worst apparent effect of these mysterious odd things is blindness to all that is divinely common. Hawthorne, I fancy, could weave one of his weird romances out of this little telepathic episode, the one strange marvel of my life, probably replacing my good old Professor by an attractive woman.

August 5. We were awakened this morning before daybreak by the furious barking of Carlo and Jack and the sound of stampeding sheep. Billy fled from his punk bed to the fire, and refused to stir into the darkness to try to gather the scattered flock, or ascertain the nature of the disturbance. It was a bear attack, as we afterward learned, and I suppose little was gained by attempting to do anything before daylight. Nevertheless, being anxious to know what was up, Carlo and I groped our way through the woods, guided by the rustling sound made by fragments of the flock, not fearing the bear, for I knew that the runaways would go from their enemy as far as possible and Carlo's nose was also to be depended upon. About half a mile east of the corral we overtook twenty or thirty of the flock and succeeded in driving them back; then turning to the westward, we traced another band of fugitives and got them back to the flock. After daybreak I discovered the remains of a sheep carcass, still warm, showing that Bruin must have been enjoying his early mutton breakfast while I was seeking the runaways. He had eaten about half of it. Six dead sheep lay in the corral, evidently smothered by the crowding and piling up of the flock against the side of the corral wall when the bear entered. Making a wide circuit of the camp, Carlo and I discovered a third band of fugitives and drove them back to camp. We also discovered another dead sheep half eaten, showing there had been two of the shaggy freebooters at this early breakfast. They were easily traced. They had each caught a sheep, jumped over the corral fence with them, carrying them as a cat carries a mouse, laid them at the foot of fir trees a hundred yards or so

back from the corral, and eaten their fill. After
breakfast I set out to seek more of the lost, and
found seventy-five at a considerable distance
from camp. In the afternoon I succeeded, with
Carlo's help, in getting them back to the flock.
I don't know whether all are together again or
not. I shall make a big fire this evening and
keep watch.

When I asked Billy why he made his bed
against the corral in rotten wood, when so many
better places offered, he replied that he "wished
to be as near the sheep as possible in case bears
should attack them." Now that the bears have
come, he has moved his bed to the far side of
the camp, and seems afraid that he may be
mistaken for a sheep.

This has been mostly a sheep day, and of
course studies have been interrupted. Nevertheless, the walk through the gloom of the
woods before the dawn was worth while, and I
have learned something about these noble bears.
Their tracks are very telling, and so are their
breakfasts. Scarce a trace of clouds to-day,
and of course our ordinary midday thunder is
wanting.

August 6. Enjoyed the grand illumination
of the camp grove, last night, from the fire we
made to frighten the bears — compensation
for loss of sleep and sheep. The noble pillars

of verdure, vividly aglow, seemed to shoot into the sky like the flames that lighted them. Nevertheless, one of the bears paid us another visit, as if more attracted than repelled by the fire, climbed into the corral, killed a sheep and made off with it without being seen, while still another was lost by trampling and suffocation against the side of the corral. Now that our mutton has been tasted, I suppose it will be difficult to put a stop to the ravages of these freebooters.

The Don arrived to-day from the lowlands with provisions and a letter. On learning the losses he had sustained, he determined to move the flock at once to the Upper Tuolumne region, saying that the bears would be sure to visit the camp every night as long as we stayed, and that no fire or noise we might make would avail to frighten them. No clouds save a few thin, lustrous touches on the eastern horizon. Thunder heard in the distance.

CHAPTER VIII

THE MONO TRAIL

August 7. Early this morning bade good-bye to the bears and blessed silver fir camp, and moved slowly eastward along the Mono Trail. At sundown camped for the night on one of the many small flowery meadows so greatly enjoyed on my excursion to Lake Tenaya. The dusty, noisy flock seems outrageously foreign and out of place in these nature gardens, more so than bears among sheep. The harm they do goes to the heart, but glorious hope lifts above all the dust and din and bids me look forward to a good time coming, when money enough will be earned to enable me to go walking where I like in pure wildness, with what I can carry on my back, and when the bread-sack is empty, run down to the nearest point on the bread-line for more. Nor will these run-downs be blanks, for, whether up or down, every step and jump on these blessed mountains is full of fine lessons.

August 8. Camp at the west end of Lake Tenaya. Arriving early, I took a walk on the glacier-polished pavements along the north

195

shore, and climbed the magnificent mountain rock at the east end of the lake, now shining in the late afternoon light. Almost every yard of its surface shows the scoring and polishing action of a great glacier that enveloped it and swept heavily over its summit, though it is about two thousand feet high above the lake and ten thousand above sea-level. This majestic, ancient ice-flood came from the eastward, as the scoring and crushing of the surface shows. Even below the waters of the lake the rock in some places is still grooved and polished; the lapping of the waves and their disintegrating action have not as yet obliterated even the superficial marks of glaciation. In climbing the steepest polished places I had to take off shoes and stockings. A fine region this for study of glacial action in mountain-making. I found many charming plants: arctic daisies, phlox, white spiræa, bryanthus, and rock-ferns, — pellæa, cheilanthes, allosorus, — fringing weathered seams all the way up to the summit; and sturdy junipers, grand old gray and brown monuments, stood bravely erect on fissured spots here and there, telling storm and avalanche stories of hundreds of winters. The view of the lake from the top is, I think, the best of all. There is another rock, more striking in form than this, standing isolated at the

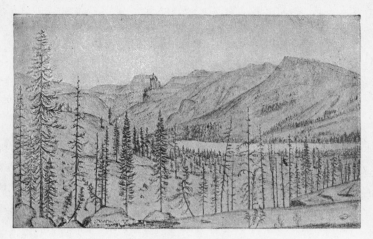

VIEW OF TENAYA LAKE SHOWING
CATHEDRAL PEAK

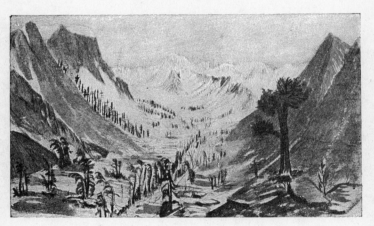

ONE OF THE TRIBUTARY FOUNTAINS OF THE
TUOLUMNE CAÑON WATERS, ON THE NORTH
SIDE OF THE HOFFMAN RANGE

head of the lake, but it is not more than half as high. It is a knob or knot of burnished granite, perhaps about a thousand feet high, apparently as flawless and strong in structure as a wave-worn pebble, and probably owes its existence to the superior resistance it offered to the action of the overflowing ice-flood.

Made sketch of the lake, and sauntered back to camp, my iron-shod shoes clanking on the pavements disturbing the chipmunks and birds. After dark went out to the shore, — not a breath of air astir, the lake a perfect mirror reflecting the sky and mountains with their stars and trees and wonderful sculpture, all their grandeur refined and doubled, — a marvelously impressive picture, that seemed to belong more to heaven than earth.

August 9. I went ahead of the flock, and crossed over the divide between the Merced and Tuolumne Basins. The gap between the east end of the Hoffman spur and the mass of mountain rocks about Cathedral Peak, though roughened by ridges and waving folds, seems to be one of the channels of a broad ancient glacier that came from the mountains on the summit of the range. In crossing this divide the ice-river made an ascent of about five hundred feet from the Tuolumne meadows. This entire region must have been overswept by ice.

MY FIRST SUMMER IN THE SIERRA

From the top of the divide, and also from the big Tuolumne Meadows, the wonderful mountain called Cathedral Peak is in sight. From every point of view it shows marked individuality. It is a majestic temple of one stone, hewn from the living rock, and adorned with spires and pinnacles in regular cathedral style. The dwarf pines on the roof look like mosses. I hope some time to climb to it to say my prayers and hear the stone sermons.

The big Tuolumne Meadows are flowery lawns, lying along the south fork of the Tuolumne River at a height of about eighty-five hundred to nine thousand feet above the sea, partially separated by forests and bars of glaciated granite. Here the mountains seem to have been cleared away or set back, so that wide-open views may be had in every direction. The upper end of the series lies at the base of Mount Lyell, the lower below the east end of the Hoffman Range, so the length must be about ten or twelve miles. They vary in width from a quarter of a mile to perhaps three quarters, and a good many branch meadows put out along the banks of the tributary streams. This is the most spacious and delightful high pleasure-ground I have yet seen. The air is keen and bracing, yet warm during the day; and though lying high in the sky, the surrounding mountains are so much higher, one feels protected as if in a grand hall. Mounts Dana and Gibbs, massive red mountains, perhaps thirteen thousand feet high or more, bound the view on the east, the Cathedral and Unicorn Peaks, with many nameless peaks, on the south, the Hoffman Range on the west, and a number of peaks unnamed, as far as I know, on the north. One of these last is much like the Cathedral. The grass of the meadows is mostly fine and silky, with exceedingly slender leaves, making a close sod, above which the panicles of minute purple flowers seem to float in airy, misty lightness, while the sod is enriched with at least three species of gentian and as many or more of orthocarpus, potentilla, ivesia, solidago, pentstemon, with their gay colors, — purple, blue, yellow, and red, — all of which I may know better ere long. A central camp will probably be made in this region, from which I hope to make long excursions into the surrounding mountains.

On the return trip I met the flock about three miles east of Lake Tenaya. Here we camped for the night near a small lake lying on top of the divide in a clump of the two-leaved pine. We are now about nine thousand feet above the sea. Small lakes abound in all sorts of situations, — on ridges, along mountain sides, and in piles of moraine boulders, most of them mere pools. Only in those cañons of the larger streams at the foot of declivities, where the down thrust of the glaciers was heaviest, do we find lakes of considerable size and depth. How grateful a task it would be to trace them all and study them! How pure their waters are, clear as crystal in polished stone basins! None of them, so far as I have seen, have fishes, I suppose on account of falls making them inaccessible. Yet one would think their eggs might get into these lakes by some chance or other; on ducks' feet, for example, or in their mouths, or in their crops, as some plant seeds are distributed. Nature has so many ways of doing such things. How did the frogs, found in all the bogs and pools and lakes, however high, manage to get up these mountains? Surely not by jumping. Such excursions through miles of dry brush and boulders would be very hard on frogs. Perhaps their stringy gelatinous spawn is occasionally entangled or glued on the feet of water birds. Anyhow, they are here and in hearty health and voice. I like their cheery tronk and crink. They take the place of songbirds at a pinch.

August 10. Another of those charming exhilarating days that make the blood dance and excite nerve currents that render one unweariable and well-nigh immortal. Had an-

other view of the broad ice-ploughed divide, and
gazed again and again at the Sierra temple and
the great red mountains east of the meadows.

We are camped near the Soda Springs on
the north side of the river. A hard time we
had getting the sheep across. They were
driven into a horseshoe bend and fairly crowded
off the bank. They seemed willing to suffer
death rather than risk getting wet, though
they swim well enough when they have to.
Why sheep should be so unreasonably afraid
of water, I don't know, but they do fear it as
soon as they are born and perhaps before. I
once saw a lamb only a few hours old approach
a shallow stream about two feet wide and an
inch deep, after it had walked only about a
hundred yards on its life journey. All the
flock to which it belonged had crossed this
inch-deep stream, and as the mother and her
lamb were the last to cross, I had a good opportunity to observe them. As soon as the
flock was out of the way, the anxious mother
crossed over and called the youngster. It
walked cautiously to the brink, gazed at the
water, bleated piteously, and refused to venture. The patient mother went back to it
again and again to encourage it, but long without avail. Like the pilgrim on Jordan's stormy
bank it feared to launch away. At length,

gathering its trembling inexperienced legs for the mighty effort, throwing up its head as if it knew all about drowning, and was anxious to keep its nose above water, it made the tremendous leap, and landed in the middle of the inch-deep stream. It seemed astonished to find that, instead of sinking over head and ears, only its toes were wet, gazed at the shining water a few seconds, and then sprang to the shore safe and dry through the dreadful adventure. All kinds of wild sheep are mountain animals, and their descendants' dread of water is not easily accounted for.

August 11. Fine shining weather, with a ten minutes' noon thunderstorm and rain. Rambling all day getting acquainted with the region north of the river. Found a small lake and many charming glacier meadows embosomed in an extensive forest of the two-leaved pine. The forest is growing on broad, almost continuous deposits of moraine material, is remarkably even in its growth, and the trees are much closer together than in any of the fir or pine woods farther down the range. The evenness of the growth would seem to indicate that the trees are all of the same age or nearly so. This regularity has probably been in great part the result of fire. I saw several large patches and strips of dead bleached spars, the ground beneath them covered with a young even growth. Fire can run in these woods, not only because the thin bark of the trees is dripping with resin, but because the growth is close, and the comparatively rich soil produces good crops of tall broad-leaved grasses on which fire can travel, even when the weather is calm. Besides these fire-killed patches there are a good many fallen uprooted trees here and there, some with the bark and needles still on, as if they had lately been blown down in some thunderstorm blast. Saw a large black-tailed deer, a buck with antlers like the upturned roots of a fallen pine.

After a long ramble through the dense encumbered woods I emerged upon a smooth meadow full of sunshine like a lake of light, about a mile and a half long, a quarter to half a mile wide, and bounded by tall arrowy pines. The sod, like that of all the glacier meadows hereabouts, is made of silky agrostis and calamagrostis chiefly; their panicles of purple flowers and purple stems, exceedingly light and airy, seem to float above the green plush of leaves like a thin misty cloud, while the sod is brightened by several species of gentian, potentilla, ivesia, orthocarpus, and their corresponding bees and butterflies. All the glacier meadows are beautiful, but few are so perfect as this one. Compared with it the most carefully leveled, licked, snipped artificial lawns of pleasure-grounds are coarse things. I should like to live here always. It is so calm and withdrawn while open to the universe in full communion with everything good. To the north of this glorious meadow I discovered the camp of some Indian hunters. Their fire was still burning, but they had not yet returned from the chase.

From meadow to meadow, every one beautiful beyond telling, and from lake to lake through groves and belts of arrowy trees, I held my way northward toward Mount Conness, finding telling beauty everywhere, while the encompassing mountains were calling "Come." Hope I may climb them all.

August 12. The sky-scenery has changed but little so far with the change in elevation. Clouds about .05. Glorious pearly cumuli tinted with purple of ineffable fineness of tone. Moved camp to the side of the glacier meadow mentioned above. To let sheep trample so divinely fine a place seems barbarous. Fortunately they prefer the succulent broad-leaved triticum and other woodland grasses to the silky species of the meadows, and therefore seldom bite them or set foot on them.

The shepherd and the Don cannot agree

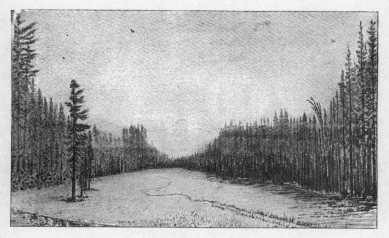

GLACIER MEADOW, ON THE HEADWATERS OF
THE TUOLUMNE, 9500 FEET ABOVE THE SEA

about methods of herding. Billy sets his dog Jack on the sheep far too often, so the Don thinks; and after some dispute to-day, in which the shepherd loudly claimed the right to dog the sheep as often as he pleased, he started for the plains. Now I suppose the care of the sheep will fall on me, though Mr. Delaney promises to do the herding himself for a while, then return to the lowlands and bring another shepherd, so as to leave me free to rove as I like.

Had another rich ramble. Pushed northward beyond the forests to the head of the general basin, where traces of glacial action are strikingly clear and interesting. The recesses among the peaks look like quarries, so raw and fresh are the moraine chips and boulders that strew the ground in Nature's glacial workshops.

Soon after my return to camp we received a visit from an Indian, probably one of the hunters whose camp I had discovered. He came from Mono, he said, with others of his tribe, to hunt deer. One that he had killed a short distance from here he was carrying on his back, its legs tied together in an ornamental bunch on his forehead. Throwing down his burden, he gazed stolidly for a few minutes in silent Indian fashion, then cut off eight or

ten pounds of venison for us, and begged a
"lill" (little) of everything he saw or could
think of — flour, bread, sugar, tobacco, whiskey, needles, etc. We gave a fair price for the
meat in flour and sugar and added a few
needles. A strangely dirty and irregular life
these dark-eyed, dark-haired, half-happy sav-
ages lead in this clean wilderness, — starvation and abundance, deathlike calm, indolence,
and admirable, indefatigable action succeed-
ing each other in stormy rhythm like winter and summer. Two things they have that
civilized toilers might well envy them — pure
air and pure water. These go far to cover and
cure the grossness of their lives. Their food is
mostly good berries, pine nuts, clover, lily
bulbs, wild sheep, antelope, deer, grouse, sage
hens, and the larvæ of ants, wasps, bees, and
other insects.

August 13. Day all sunshine, dawn and
evening purple, noon gold, no clouds, air motionless. Mr. Delaney arrived with two shepherds, one of them an Indian. On his way up
from the plains he left some provisions at the
Portuguese camp on Porcupine Creek near
our old Yosemite camp, and I set out this
morning with one of the pack animals to fetch
them. Arrived at the Porcupine camp at noon,
and might have returned to the Tuolumne late

in the evening, but concluded to stay over night with the Portuguese shepherds at their pressing invitation. They had sad stories to tell of losses from the Yosemite bears, and were so discouraged they seemed on the point of leaving the mountains; for the bears came every night and helped themselves to one or several of the flock in spite of all their efforts to keep them off.

I spent the afternoon in a grand ramble along the Yosemite walls. From the highest of the rocks called the Three Brothers, I enjoyed a magnificent view comprehending all the upper half of the floor of the valley and nearly all the rocks of the walls on both sides and at the head, with snowy peaks in the background. Saw also the Vernal and Nevada Falls, a truly glorious picture, — rocky strength and permanence combined with beauty of plants frail and fine and evanescent; water descending in thunder, and the same water gliding through meadows and groves in gentlest beauty. This standpoint is about eight thousand feet above the sea, or four thousand feet above the floor of the valley, and every tree, though looking small and feathery, stands in admirable clearness, and the shadows they cast are as distinct in outline as if seen at a distance of a few yards. They appeared even more so. No words will ever describe the exquisite beauty and charm of this mountain park — Nature's landscape garden at once tenderly beautiful and sublime. No wonder it draws nature-lovers from all over the world.

Glacial action even on this lofty summit is plainly displayed. Not only has all the lovely valley now smiling in sunshine been filled to the brim with ice, but it has been deeply overflowed.

I visited our old Yosemite camp-ground on the head of Indian Creek, and found it fairly patted and smoothed down with bear-tracks. The bears had eaten all the sheep that were smothered in the corral, and some of the grand animals must have died, for Mr. Delaney, before leaving camp, put a large quantity of poison in the carcasses. All sheep-men carry strychnine to kill coyotes, bears, and panthers, though neither coyotes nor panthers are at all numerous in the upper mountains. The little dog-like wolves are far more numerous in the foothill region and on the plains, where they find a better supply of food, — saw only one panther-track above eight thousand feet.

On my return after sunset to the Portuguese camp I found the shepherds greatly excited over the behavior of the bears that have learned to like mutton. "They are getting

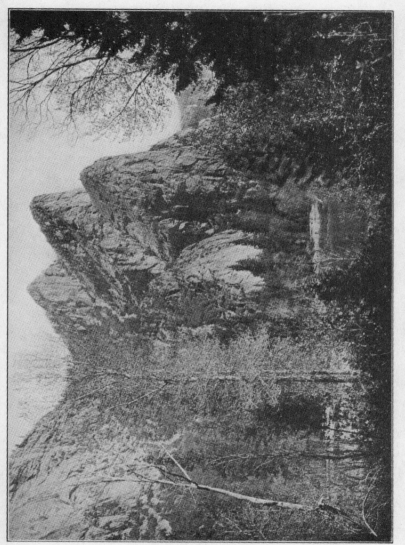

THE THREE BROTHERS, YOSEMITE NATIONAL PARK

worse and worse," they lamented. Not willing to wait decently until after dark for their
suppers, they come and kill and eat their fill
in broad daylight. The evening before my
arrival, when the two shepherds were leisurely
driving the flock toward camp half an hour
before sunset, a hungry bear came out of the
chaparral within a few yards of them and
shuffled deliberately toward the flock. "Portuguese Joe," who always carried a gun loaded
with buckshot, fired excitedly, threw down
his gun, fled to the nearest suitable tree, and
climbed to a safe height without waiting to
see the effect of his shot. His companion also
ran, but said that he saw the bear rise on its
hind legs and throw out its arms as if feeling
for somebody, and then go into the brush as if
wounded.

At another of their camps in this neighborhood, a bear with two cubs attacked the flock
before sunset, just as they were approaching
the corral. Joe promptly climbed a tree out
of danger, while Antone, rebuking his companion for cowardice in abandoning his charge,
said that he was not going to let bears "eat up
his sheeps" in daylight, and rushed towards
the bears, shouting and setting his dog on
them. The frightened cubs climbed a tree,
but the mother ran to meet the shepherd and

seemed anxious to fight. Antone stood astonished for a moment, eyeing the oncoming bear, then turned and fled, closely pursued. Unable to reach a suitable tree for climbing, he ran to the camp and scrambled up to the roof of the little cabin; the bear followed, but did not climb to the roof, — only stood glaring up at him for a few minutes, threatening him and holding him in mortal terror, then went to her cubs, called them down, went to the flock, caught a sheep for supper, and vanished in the brush. As soon as the bear left the cabin, the trembling Antone begged Joe to show him a good safe tree, up which he climbed like a sailor climbing a mast, and remained as long as he could hold on, the tree being almost branchless. After these disastrous experiences the two shepherds chopped and gathered large piles of dry wood and made a ring of fire around the corral every night, while one with a gun kept watch from a comfortable stage built on a neighboring pine that commanded a view of the corral. This evening the show made by the circle of fire was very fine, bringing out the surrounding trees in most impressive relief, and making the thousands of sheep eyes glow like a glorious bed of diamonds.

August 14. Up to the time I went to bed

last night all was quiet, though we expected the shaggy freebooters every minute. They did not come till near midnight, when a pair walked boldly to the corral between two of the great fires, climbed in, killed two sheep and smothered ten, while the frightened watcher in the tree did not fire a single shot, saying that he was afraid he might kill some of the sheep, for the bears got into the corral before he got a good clear view of them. I told the shepherds they should at once move the flock to another camp. "Oh, no use, no use," they lamented; "where we go, the bears go too. See my poor dead sheeps — soon all dead. No use try another camp. We go down to the plains." And as I afterwards learned, they were driven out of the mountains a month before the usual time. Were bears much more numerous and destructive, the sheep would be kept away altogether.

It seems strange that bears, so fond of all sorts of flesh, running the risks of guns and fires and poison, should never attack men except in defense of their young. How easily and safely a bear could pick us up as we lie asleep! Only wolves and tigers seem to have learned to hunt man for food, and perhaps sharks and crocodiles. Mosquitoes and other insects would, I suppose, devour a helpless man in some parts of the world, and so might lions, leopards, wolves, hyenas, and panthers at times if pressed by hunger, — but under ordinary circumstances, perhaps, only the tiger among land animals may be said to be a man-eater, — unless we add man himself.

Clouds as usual about .05. Another glorious Sierra day, warm, crisp, fragrant, and clear. Many of the flowering plants have gone to seed, but many others are unfolding their petals every day, and the firs and pines are more fragrant than ever. Their seeds are nearly ripe, and will soon be flying in the merriest flocks that ever spread a wing.

On the way back to our Tuolumne camp, I enjoyed the scenery if possible more than when it first came to view. Every feature already seems familiar as if I had lived here always. I never weary gazing at the wonderful Cathedral. It has more individual character than any other rock or mountain I ever saw, excepting perhaps the Yosemite South Dome. The forests, too, seem kindly familiar, and the lakes and meadows and glad singing streams. I should like to dwell with them forever. Here with bread and water I should be content. Even if not allowed to roam and climb, tethered to a stake or tree in some meadow or grove, even then I should be content forever. Bathed in such beauty, watching the expressions ever varying on the faces of the mountains, watching the stars, which here have a glory that the lowlander never dreams of, watching the circling seasons, listening to the songs of the waters and winds and birds, would be endless pleasure. And what glorious cloudlands I should see, storms and calms, — a new heaven and a new earth every day, aye and new inhabitants. And how many visitors I should have. I feel sure I should not have one dull moment. And why should this appear extravagant? It is only common sense, a sign of health, genuine, natural, all-awake health. One would be at an endless Godful play, and what speeches and music and acting and scenery and lights! — sun, moon, stars, auroras. Creation just beginning, the morning stars "still singing together and all the sons of God shouting for joy."

CHAPTER IX

BLOODY CAÑON AND MONO LAKE

August 21. Have just returned from a fine wild excursion across the range to Mono Lake, by way of the Mono or Bloody Cañon Pass. Mr. Delaney has been good to me all summer, lending a helping, sympathizing hand at every opportunity, as if my wild notions and rambles and studies were his own. He is one of those remarkable California men who have been overflowed and denuded and remodeled by the excitements of the gold fields, like the Sierra landscapes by grinding ice, bringing the harder bosses and ridges of character into relief, — a tall, lean, big-boned, big-hearted Irishman, educated for a priest in Maynooth College, — lots of good in him, shining out now and then in this mountain light. Recognizing my love of wild places, he told me one evening that I ought to go through Bloody Cañon, for he was sure I should find it wild enough. He had not been there himself, he said, but had heard many of his mining friends speak of it as the wildest of all the Sierra passes. Of course I was glad to go. It lies just

214

to the east of our camp and swoops down from the summit of the range to the edge of the Mono Desert, making a descent of about four thousand feet in a distance of about four miles. It was known and traveled as a pass by wild animals and the Indians long before its discovery by white men in the gold year of 1858, as is shown by old trails which come together at the head of it. The name may have been suggested by the red color of the metamorphic slates in which the cañon abounds, or by the blood stains on the rocks from the unfortunate animals that were compelled to slide and shuffle over the sharp-angled boulders.

Early in the morning I tied my notebook and some bread to my belt, and strode away full of eager hope, feeling that I was going to have a glorious revel. The glacier meadows that lay along my way served to soothe my morning speed, for the sod was full of blue gentians and daisies, kalmia and dwarf vaccinium, calling for recognition as old friends, and I had to stop many times to examine the shining rocks over which the ancient glacier had passed with tremendous pressure, polishing them so well that they reflected the sunlight like glass in some places, while fine striæ, seen clearly through a lens, indicated the direction in which the ice had flowed. On some of the sloping polished pavements abrupt steps occur, showing that occasionally large masses of the rock had given way before the glacial pressure, as well as small particles; moraines, too, some scattered, others regular like long curving embankments and dams, occur here and there, giving the general surface of the region a young, new-made appearance. I watched the gradual dwarfing of the pines as I ascended, and the corresponding dwarfing of nearly all the rest of the vegetation. On the slopes of Mammoth Mountain, to the south of the pass, I saw many gaps in the woods reaching from the upper edge of the timber-line down to the level meadows, where avalanches of snow had descended, sweeping away every tree in their paths as well as the soil they were growing in, leaving the bed-rock bare. The trees are nearly all uprooted, but a few that had been extremely well anchored in clefts of the rock were broken off near the ground. It seems strange at first sight that trees that had been allowed to grow for a century or more undisturbed should in their old age be thus swished away at a stroke. Such avalanches can only occur under rare conditions of weather and snowfall. No doubt on some positions of the mountain slopes the inclination and smoothness of the surface is such that avalanches must occur every winter, or even after every heavy snowstorm, and of course no trees or even bushes can grow in their channels. I noticed a few clean-swept slopes of this kind. The uprooted trees that had grown in the pathway of what might be called "century avalanches" were piled in windrows, and tucked snugly against the wall-trees of the gaps, heads downward, excepting a few that were carried out into the open ground of the meadows, where the heads of the avalanches had stopped. Young pines, mostly the two-leaved and the white-barked, are already springing up in these cleared gaps. It would be interesting to ascertain the age of these saplings, for thus we should gain a fair approximation to the year that the great avalanches occurred. Perhaps most or all of them occurred the same winter. How glad I should be if free to pursue such studies!

Near the summit at the head of the pass I found a species of dwarf willow lying perfectly flat on the ground, making a nice, soft, silky gray carpet, not a single stem or branch more than three inches high; but the catkins, which are now nearly ripe, stand erect and make a close, nearly regular gray growth, being larger than all the rest of the plants. Some of these interesting dwarfs have only one catkin —

willow bushes reduced to their lowest terms.
I found patches of dwarf vaccinium also forming smooth carpets, closely pressed to the
ground or against the sides of stones, and covered with round pink flowers in lavish abundance as if they had fallen from the sky like
hail. A little higher, almost at the very head
of the pass, I found the blue arctic daisy and
purple-flowered bryanthus, the mountain's
own darlings, gentle mountaineers face to face
with the sky, kept safe and warm by a thousand miracles, seeming always the finer and
purer the wilder and stormier their homes.
The trees, tough and resiny, seem unable to
go a step farther; but up and up, far above the
tree-line, these tender plants climb, cheerily
spreading their gray and pink carpets right
up to the very edges of the snow-banks in deep
hollows and shadows. Here, too, is the familiar
robin, tripping on the flowery lawns, bravely
singing the same cheery song I first heard
when a boy in Wisconsin newly arrived from
old Scotland. In this fine company sauntering enchanted, taking no heed of time, I at
length entered the gate of the pass, and the
huge rocks began to close around me in all
their mysterious impressiveness. Just then I
was startled by a lot of queer, hairy, muffled
creatures coming shuffling, shambling, wallow-

ing toward me as if they had no bones in their bodies. Had I discovered them while they were yet a good way off, I should have tried to avoid them. What a picture they made contrasted with the others I had just been admiring. When I came up to them, I found that they were only a band of Indians from Mono on their way to Yosemite for a load of acorns. They were wrapped in blankets made of the skins of sage-rabbits. The dirt on some of the faces seemed almost old enough and thick enough to have a geological significance; some were strangely blurred and divided into sections by seams and wrinkles that looked like cleavage joints, and had a worn abraded look as if they had lain exposed to the weather for ages. I tried to pass them without stopping, but they would n't let me; forming a dismal circle about me, I was closely besieged while they begged whiskey or tobacco, and it was hard to convince them that I had n't any. How glad I was to get away from the gray, grim crowd and see them vanish down the trail! Yet it seems sad to feel such desperate repulsion from one's fellow beings, however degraded. To prefer the society of squirrels and woodchucks to that of our own species must surely be unnatural. So with a fresh breeze and a hill or mountain between us I

must wish them Godspeed and try to pray and
sing with Burns, "It's coming yet, for a' that,
that man to man, the warld o'er, shall brothers
be for a' that."

How the day passed I hardly know. By
the map I have come only about ten or twelve
miles, though the sun is already low in the
west, showing how long I must have lingered,
observing, sketching, taking notes among the
glaciated rocks and moraines and Alpine
flower-beds.

At sundown the somber crags and peaks
were inspired with the ineffable beauty of the
alpenglow, and a solemn, awful stillness hushed
everything in the landscape. Then I crept
into a hollow by the side of a small lake near
the head of the cañon, smoothed a sheltered
spot, and gathered a few pine tassels for a bed.
After the short twilight began to fade I kin-
dled a sunny fire, made a tin cupful of tea, and
lay down to watch the stars. Soon the night-
wind began to flow from the snowy peaks overhead, at first only a gentle breathing, then
gaining strength, in less than an hour rum-
bled in massive volume something like a boisterous stream in a boulder-choked channel,
roaring and moaning down the cañon as if the
work it had to do was tremendously important and fateful; and mingled with these storm

tones were those of the waterfalls on the north side of the cañon, now sounding distinctly, now smothered by the heavier cataracts of air, making a glorious psalm of savage wildness. My fire squirmed and struggled as if ill at ease, for though in a sheltered nook, detached masses of icy wind often fell like icebergs on top of it, scattering sparks and coals, so that I had to keep well back to avoid being burned. But the big resiny roots and knots of the dwarf pine could neither be beaten out nor blown away, and the flames, now rushing up in long lances, now flattened and twisted on the rocky ground, roared as if trying to tell the storm stories of the trees they belonged to, as the light given out was telling the story of the sunshine they had gathered in centuries of summers.

The stars shone clear in the strip of sky between the huge dark cliffs; and as I lay recalling the lessons of the day, suddenly the full moon looked down over the cañon wall, her face apparently filled with eager concern, which had a startling effect, as if she had left her place in the sky and had come down to gaze on me alone, like a person entering one's bedroom. It was hard to realize that she was in her place in the sky, and was looking abroad on half the globe, land and sea, mountains,

plains, lakes, rivers, oceans, ships, cities with
their myriads of inhabitants sleeping and
waking, sick and well. No, she seemed to be
just on the rim of Bloody Cañon and looking
only at me. This was indeed getting near to
Nature. I remember watching the harvest
moon rising above the oak trees in Wisconsin
apparently as big as a cart-wheel and not
farther than half a mile distant. With these
exceptions I might say I never before had seen
the moon, and this night she seemed so full
of life and so near, the effect was marvelously
impressive and made me forget the Indians,
the great black rocks above me, and the wild
uproar of the winds and waters making their
way down the huge jagged gorge. Of course
I slept but little and gladly welcomed the
dawn over the Mono Desert. By the time I
had made a cupful of tea the sunbeams were
pouring through the cañon, and I set forth,
gazing eagerly at the tremendous walls of red
slates savagely hacked and scarred and apparently ready to fall in avalanches great enough
to choke the pass and fill up the chain of lakelets. But soon its beauties came to view, and I
bounded lightly from rock to rock, admiring
the polished bosses shining in the slant sunshine
with glorious effect in the general roughness of
moraines and avalanche taluses, even toward

the head of the cañon near the highest fountains of the ice. Here, too, are most of the lowly plant people seen yesterday on the other side of the divide now opening their beautiful eyes. None could fail to glory in Nature's tender care for them in so wild a place. The little ouzel is flitting from rock to rock along the rapid swirling Cañon Creek, diving for breakfast in icy pools, and merrily singing as if the huge rugged avalanche-swept gorge was the most delightful of all its mountain homes. Besides a high fall on the north wall of the cañon, apparently coming direct from the sky, there are many narrow cascades, bright silvery ribbons zigzagging down the red cliffs, tracing the diagonal cleavage joints of the metamorphic slates, now contracted and out of sight, now leaping from ledge to ledge in filmy sheets through which the sunbeams sift. And on the main Cañon Creek, to which all these are tributary, is a series of small falls, cascades, and rapids extending all the way down to the foot of the cañon, interrupted only by the lakes in which the tossed and beaten waters rest. One of the finest of the cascades is outspread on the face of a precipice, its waters separated into ribbon-like strips, and woven into a diamond-like pattern by tracing the cleavage joints of the rock, while tufts of bryanthus, grass, sedge, saxifrage form beautiful fringes. Who could imagine beauty so fine in so savage a place? Gardens are blooming in all sorts of nooks and hollows, — at the head alpine eriogonums, erigerons, saxifrages, gentians, cowania, bush primula; in the middle region larkspur, columbine, orthocarpus, castilleia, harebell, epilobium, violets, mints, yarrow; near the foot sunflowers, lilies, brier rose, iris, lonicera, clematis.

One of the smallest of the cascades, which I name the Bower Cascade, is in the lower region of the pass, where the vegetation is snowy and luxuriant. Wild rose and dogwood form dense masses overarching the stream, and out of this bower the creek, grown strong with many indashing tributaries, leaps forth into the light, and descends in a fluted curve thick-sown with crisp flashing spray. At the foot of the cañon there is a lake formed in part at least by the damming of the stream by a terminal moraine. The three other lakes in the cañon are in basins eroded from the solid rock, where the pressure of the glacier was greatest, and the most resisting portions of the basin rims are beautifully, tellingly polished. Below Moraine Lake at the foot of the cañon there are several old lake-basins lying between the large lateral moraines which extend out into the desert. These basins are now completely filled up by the material carried in by the streams, and changed to dry sandy flats covered mostly by grass and artemisia and sun-loving flowers. All these lower lake-basins were evidently formed by terminal moraine dams deposited where the receding glacier had lingered during short periods of less waste, or greater snowfall, or both.

Looking up the cañon from the warm sunny edge of the Mono plain my morning ramble seems a dream, so great is the change in the vegetation and climate. The lilies on the bank of Moraine Lake are higher than my head, and the sunshine is hot enough for palms. Yet the snow round the arctic gardens at the summit of the pass is plainly visible, only about four miles away, and between lie specimen zones of all the principal climates of the globe. In little more than an hour one may swoop down from winter to summer, from an Arctic to a torrid region, through as great changes of climate as one would encounter in traveling from Labrador to Florida.

The Indians I had met near the head of the cañon had camped at the foot of it the night before they made the ascent, and I found their fire still smoking on the side of a small tributary

stream near Moraine Lake; and on the edge of
what is called the Mono Desert, four or five
miles from the lake, I came to a patch of elymus, or wild rye, growing in magnificent waving clumps six or eight feet high, bearing heads
six to eight inches long. The crop was ripe, and
Indian women were gathering the grain in
baskets by bending down large handfuls, beating out the seed, and fanning it in the wind.
The grains are about five eighths of an inch
long, dark-colored and sweet. I fancy the bread
made from it must be as good as wheat bread.
A fine squirrelish employment this wild grain
gathering seems, and the women were evidently
enjoying it, laughing and chattering and looking almost natural, though most Indians I have
seen are not a whit more natural in their lives
than we civilized whites. Perhaps if I knew
them better I should like them better. The
worst thing about them is their uncleanliness.
Nothing truly wild is unclean. Down on the
shore of Mono Lake I saw a number of their
flimsy huts on the banks of streams that dash
swiftly into that dead sea, — mere brush tents
where they lie and eat at their ease. Some of
the men were feasting on buffalo berries, lying
beneath the tall bushes now red with fruit. The
berries are rather insipid, but they must needs
be wholesome, since for days and weeks the In-

dians, it is said, eat nothing else. In the season they in like manner depend chiefly on the fat larvæ of a fly that breeds in the salt water of the lake, or on the big fat corrugated caterpillars of a species of silkworm that feeds on the leaves of the yellow pine. Occasionally a grand rabbit-drive is organized and hundreds are slain with clubs on the lake shore, chased and frightened into a dense crowd by dogs, boys, girls, men and women, and rings of sage brush fire, when of course they are quickly killed. The skins are made into blankets. In the autumn the more enterprising of the hunters bring in a good many deer, and rarely a wild sheep from the high peaks. Antelopes used to be abundant on the desert at the base of the interior mountain-ranges. Sage hens, grouse, and squirrels help to vary their wild diet of worms; pine nuts also from the small interesting *Pinus monophylla*, and good bread and good mush are made from acorns and wild rye. Strange to say, they seem to like the lake larvæ best of all. Long windrows are washed up on the shore, which they gather and dry like grain for winter use. It is said that wars, on account of encroachments on each other's worm-grounds, are of common occurrence among the various tribes and families. Each claims a certain marked portion of the shore.

The pine nuts are delicious — large quantities are gathered every autumn. The tribes of the west flank of the range trade acorns for worms and pine nuts. The squaws carry immense loads on their backs across the rough passes and down the range, making journeys of about forty or fifty miles each way.

The desert around the lake is surprisingly flowery. In many places among the sage bushes I saw mentzelia, abronia, aster, bigelovia, and gilia, all of which seemed to enjoy the hot sunshine. The abronia, in particular, is a delicate, fragrant, and most charming plant.

Opposite the mouth of the cañon a range of volcanic cones extends southward from the lake, rising abruptly out of the desert like a chain of mountains. The largest of the cones are about twenty-five hundred feet high above the lake level, have well-formed craters, and all of them are evidently comparatively recent additions to the landscape. At a distance of a few miles they look like heaps of loose ashes that have never been blest by either rain or snow, but, for a' that and a' that, yellow pines are climbing their gray slopes, trying to clothe them and give beauty for ashes. A country of wonderful contrasts. Hot deserts bounded by snow-laden mountains, — cinders and ashes scattered on glacier-polished pavements, —

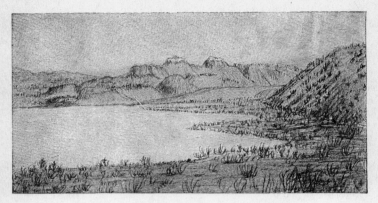

MONO LAKE AND VOLCANIC CONES,
LOOKING SOUTH

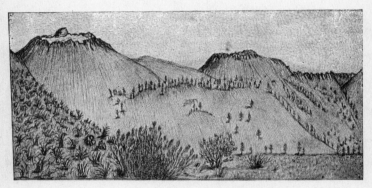

HIGHEST MONO VOLCANIC CONES (NEAR VIEW)

frost and fire working together in the making of beauty. In the lake are several volcanic islands, which show that the waters were once mingled with fire.

Glad to get back to the green side of the mountains, though I have greatly enjoyed the gray east side and hope to see more of it. Reading these grand mountain manuscripts displayed through every vicissitude of heat and cold, calm and storm, upheaving volcanoes and down-grinding glaciers, we see that everything in Nature called destruction must be creation — a change from beauty to beauty.

Our glacier meadow camp north of the Soda Springs seems more beautiful every day. The grass covers all the ground though the leaves are thread-like in fineness, and in walking on the sod it seems like a plush carpet of marvelous richness and softness, and the purple panicles brushing against one's feet are not felt. This is a typical glacier meadow, occupying the basin of a vanished lake, very definitely bounded by walls of the arrowy two-leaved pines drawn up in a handsome orderly array like soldiers on parade. There are many other meadows of the same kind hereabouts imbedded in the woods. The main big meadows along the river are the same in general and extend with but little interruption for ten or

twelve miles, but none I have seen are so finely
finished and perfect as this one. It is richer in
flowering plants than the prairies of Wisconsin
and Illinois were when in all their wild glory.
The showy flowers are mostly three species of
gentian, a purple and yellow orthocarpus, a
golden-rod or two, a small blue pentstemon
almost like a gentian, potentilla, ivesia, pedi-
cularis, white violet, kalmia, and bryanthus.
There are no coarse weedy plants. Through
this flowery lawn flows a stream silently gliding, swirling, slipping as if careful not to make
the slightest noise. It is only about three feet
wide in most places, widening here and there
into pools six or eight feet in diameter with no
apparent current, the banks bossily rounded by
the down-curving mossy sod, grass panicles
over-leaning like miniature pine trees, and rugs
of bryanthus spreading here and there over
sunken boulders. At the foot of the meadow the
stream, rich with the juices of the plants it has
refreshed, sings merrily down over shelving rock
ledges on its way to the Tuolumne River. The
sublime, massive Mount Dana and its companions, green, red, and white, loom impressively
above the pines along the eastern horizon; a
range or spur of gray rugged granite crags and
mountains on the north; the curiously crested
and battlemented Mount Hoffman on the west;

and the Cathedral Range on the south with its grand Cathedral Peak, Cathedral Spires, Unicorn Peak, and several others, gray and pointed or massively rounded.

CHAPTER X

THE TUOLUMNE CAMP

August 22. Clouds none, cool west wind, slight hoarfrost on the meadows. Carlo is missing; have been seeking him all day. In the thick woods between camp and the river, among tall grass and fallen pines, I discovered a baby fawn. At first it seemed inclined to come to me; but when I tried to catch it, and got within a rod or two, it turned and walked softly away, choosing its steps like a cautious, stealthy, hunting cat. Then, as if suddenly called or alarmed, it began to buck and run like a grown deer, jumping high above the fallen trunks, and was soon out of sight. Possibly its mother may have called it, but I did not hear her. I don't think fawns ever leave the home thicket or follow their mothers until they are called or frightened. I am distressed about Carlo. There are several other camps and dogs not many miles from here, and I still hope to find him. He never left me before. Panthers are very rare here, and I don't think any of these cats would dare touch him. He knows bears too well to be caught by them, and as for Indians, they don't want him.

232

THE TUOLUMNE CAMP

August 23. Cool, bright day, hinting Indian summer. Mr. Delaney has gone to the Smith Ranch, on the Tuolumne below Hetch-Hetchy Valley, thirty-five or forty miles from here, so I'll be alone for a week or more, — not really alone, for Carlo has come back. He was at a camp a few miles to the northwestward. He looked sheepish and ashamed when I asked him where he had been and why he had gone away without leave. He is now trying to get me to caress him and show signs of forgiveness. A wondrous wise dog. A great load is off my mind. I could not have left the mountains without him. He seems very glad to get back to me.

Rose and crimson sunset, and soon after the stars appeared the moon rose in most impressive majesty over the top of Mount Dana. I sauntered up the meadow in the white light. The jet-black tree-shadows were so wonderfully distinct and substantial looking, I often stepped high in crossing them, taking them for black charred logs.

August 24. Another charming day, warm and calm soon after sunrise, clouds only about .01, — faint, silky cirrus wisps, scarcely visible. Slight frost, Indian summerish, the mountains growing softer in outline and dreamy looking, their rough angles melted off, apparently. Sky at evening with fine, dark, subdued purple, almost like the evening purple of the San Joaquin plains in settled weather. The moon is now gazing over the summit of Dana. Glorious exhilarating air. I wonder if in all the world there is another mountain range of equal height blessed with weather so fine, and so openly kind and hospitable and approachable.

August 25. Cool as usual in the morning, quickly changing to the ordinary serene generous warmth and brightness. Toward evening the west wind was cool and sent us to the campfire. Of all Nature's flowery carpeted mountain halls none can be finer than this glacier meadow. Bees and butterflies seem as abundant as ever. The birds are still here, showing no sign of leaving for winter quarters though the frost must bring them to mind. For my part I should like to stay here all winter or all my life or even all eternity.

August 26. Frost this morning; all the meadow grass and some of the pine needles sparkling with irised crystals, — flowers of light. Large picturesque clouds, craggy like rocks, are piled on Mount Dana, reddish in color like the mountain itself; the sky for a few degrees around the horizon is pale purple, into which the pines dip their spires with fine effect. Spent the day as usual looking about me, watching the changing lights, the ripening autumn colors of the grass, seeds, late-blooming gentians, asters, goldenrods; parting the meadow grass here and there and looking down into the underworld of mosses and liverworts; watching the busy ants and beetles and other small people at work and play like squirrels and bears in a forest; studying the formation of lakes and meadows, moraines, mountain sculpture; making small beginnings in these directions, charmed by the serene beauty of everything.

The day has been extra cloudy, though bright on the whole, for the clouds were brighter than common. Clouds about .15, which in Switzerland would be considered extra clear. Probably more free sunshine falls on this majestic range than on any other in the world I've ever seen or heard of. It has the brightest weather, brightest glacier-polished rocks, the greatest abundance of irised spray from its glorious waterfalls, the brightest forests of silver firs and silver pines, more starshine, moonshine, and perhaps more crystalshine than any other mountain chain, and its countless mirror lakes, having more light poured into them, glow and spangle most. And how glorious the shining after the short summer showers and after frosty nights when the morning sunbeams are pouring through the crystals on the grass and pine needles, and how ineffably spiritually fine is the morning-glow on the mountain-tops and the alpenglow of evening. Well may the Sierra be named, not the Snowy Range, but the Range of Light.

August 27. Clouds only .05, — mostly white and pink cumuli over the Hoffman spur towards evening, — frosty morning. Crystals grow in marvelous beauty and perfection of form these still nights, every one built as carefully as the grandest holiest temple, as if planned to endure forever.

Contemplating the lace-like fabric of streams outspread over the mountains, we are reminded that everything is flowing — going somewhere, animals and so-called lifeless rocks as well as water. Thus the snow flows fast or slow in grand beauty-making glaciers and avalanches; the air in majestic floods carrying minerals, plant leaves, seeds, spores, with streams of music and fragrance; water streams carrying rocks both in solution and in the form of mud particles, sand, pebbles, and boulders. Rocks flow from volcanoes like water from springs, and animals flock together and flow in currents modified by stepping, leaping, gliding, flying, swimming, etc. While the stars go streaming through space pulsed on and on forever like blood globules in Nature's warm heart.

August 28. The dawn a glorious song of

color. Sky absolutely cloudless. A fine crop of hoarfrost. Warm after ten o'clock. The gentians don't mind the first frost though their petals seem so delicate; they close every night as if going to sleep, and awake fresh as ever in the morning sun-glory. The grass is a shade browner since last week, but there are no nipped wilted plants of any sort as far as I have seen. Butterflies and the grand host of smaller flies are benumbed every night, but they hover and dance in the sunbeams over the meadows before noon with no apparent lack of playful, joyful life. Soon they must all fall like petals in an orchard, dry and wrinkled, not a wing of all the mighty host left to tingle the air. Nevertheless new myriads will arise in the spring, rejoicing, exulting, as if laughing cold death to scorn.

August 29. Clouds about .05, slight frost. Bland serene Indian summer weather. Have been gazing all day at the mountains, watching the changing lights. More and more plainly are they clothed with light as a garment, white tinged with pale purple, palest during the midday hours, richest in the morning and evening. Everything seems consciously peaceful, thoughtful, faithfully waiting God's will.

August 30. This day just like yesterday. A few clouds motionless and apparently with **no** work to do beyond looking beautiful. Frost enough for crystal building, — glorious fields of ice-diamonds destined to last but a night. How lavish is Nature building, pulling down, creating, destroying, chasing every material particle from form to form, ever changing, ever beautiful.

Mr. Delaney arrived this morning. Felt not a trace of loneliness while he was gone. On the contrary, I never enjoyed grander company. The whole wilderness seems to be alive and familiar, full of humanity. The very stones seem talkative, sympathetic, brotherly. No wonder when we consider that we all have the same Father and Mother.

August 31. Clouds .05. Silky cirrus wisps and fringes so fine they almost escape notice. Frost enough for another crop of crystals on the meadows but none on the forests. The gentians, goldenrods, asters, etc., don't seem to feel it; neither petals nor leaves are touched though they seem so tender. Every day opens and closes like a flower, noiseless, effortless. Divine peace glows on all the majestic landscape like the silent enthusiastic joy that sometimes transfigures a noble human face.

September 1. Clouds .05 — motionless, of no particular color — ornaments with no hint of rain or snow in them. Day all calm — another grand throb of Nature's heart, ripening late flowers and seeds for next summer, full of life and the thoughts and plans of life to come, and full of ripe and ready death beautiful as life, telling divine wisdom and goodness and immortality. Have been up Mount Dana, making haste to see as much as I can now that the time of departure is drawing nigh. The views from the summit reach far and wide, eastward over the Mono Lake and Desert; mountains beyond mountains looking strangely barren and gray and bare like heaps of ashes dumped from the sky. The lake, eight or ten miles in diameter, shines like a burnished disk of silver, no trees about its gray, ashy, cindery shores. Looking westward, the glorious forests are seen sweeping over countless ridges and hills, girdling domes and subordinate mountains, fringing in long curving lines the dividing ridges, and filling every hollow where the glaciers have spread soil-beds however rocky or smooth. Looking northward and southward along the axis of the range, you see the glorious array of high mountains, crags and peaks and snow, the fountain-heads of rivers that are flowing west to the sea through the famous Golden Gate, and east to hot salt lakes and deserts to evaporate and hurry back into the sky. Innumerable lakes are shining like eyes beneath heavy rock brows, bare or tree fringed, or imbedded in black forests. Meadow openings in the woods seem as numerous as the lakes or perhaps more so. Far up the moraine-covered slopes and among crumbling rocks I found many delicate hardy plants, some of them still in flower. The best gains of this trip were the lessons of unity and interrelation of all the features of the landscape revealed in general views. The lakes and meadows are located just where the ancient glaciers bore heaviest at the foot of the steepest parts of their channels, and of course their longest diameters are approximately parallel with each other and with the belts of forests growing in long curving lines on the lateral and medial moraines, and in broad outspreading fields on the terminal beds deposited toward the end of the ice period when the glaciers were receding. The domes, ridges, and spurs also show the influence of glacial action in their forms, which approximately seem to be the forms of greatest strength with reference to the stress of oversweeping, past-sweeping, down-grinding ice-streams; survivals of the most resisting masses, or those most favorably situated. How interesting everything is! Every rock, mountain, stream, plant, lake, lawn, forest, garden, bird, beast, insect seems

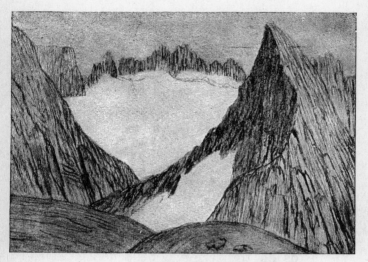

ONE OF THE HIGHEST MOUNT RITTER
FOUNTAINS

to call and invite us to come and learn something of its history and relationship. But shall the poor ignorant scholar be allowed to try the lessons they offer? It seems too great and good to be true. Soon I'll be going to the lowlands. The bread camp must soon be removed. If I had a few sacks of flour, an axe, and some matches, I would build a cabin of pine logs, pile up plenty of firewood about it and stay all winter to see the grand fertile snow-storms, watch the birds and animals that winter thus high, how they live, how the forests look snow-laden or buried, and how the avalanches look and sound on their way down the mountains. But now I'll have to go, for there is nothing to spare in the way of provisions. I'll surely be back, however, surely I'll be back. No other place has ever so overwhelmingly attracted me as this hospitable, Godful wilderness.

September 2. A grand, red, rosy, crimson day, — a perfect glory of a day. What it means I don't know. It is the first marked change from tranquil sunshine with purple mornings and evenings and still, white noons. There is nothing like a storm, however. The average cloudiness only about .08, and there is no sighing in the woods to betoken a big weather change. The sky was red in the morning and evening, the color not diffused like the ordinary purple glow, but loaded upon separate well-defined clouds that remained motionless, as if anchored around the jagged mountain-fenced horizon. A deep-red cap, bluffy around its sides, lingered a long time on Mount Dana and Mount Gibbs, drooping so low as to hide most of their bases, but leaving Dana's round summit free, which seemed to float separate and alone over the big crimson cloud. Mammoth Mountain, to the south of Gibbs and Bloody Cañon, striped and spotted with snow-banks and clumps of dwarf pine, was also favored with a glorious crimson cap, in the making of which there was no trace of economy — a huge bossy pile colored with a perfect passion of crimson that seemed important enough to be sent off to burn among the stars in majestic independence. One is constantly reminded of the infinite lavishness and fertility of Nature — inexhaustible abundance amid what seems enormous waste. And yet when we look into any of her operations that lie within reach of our minds, we learn that no particle of her material is wasted or worn out. It is eternally flowing from use to use, beauty to yet higher beauty; and we soon cease to lament waste and death, and rather rejoice and exult in the imperishable, unspendable

wealth of the universe, and faithfully watch
and wait the reappearance of everything that
melts and fades and dies about us, feeling sure
that its next appearance will be better and
more beautiful than the last.

I watched the growth of these red-lands of
the sky as eagerly as if new mountain ranges
were being built. Soon the group of snowy
peaks in whose recesses lie the highest fountains of the Tuolumne, Merced, and North
Fork of the San Joaquin were decorated with
majestic colored clouds like those already described, but more complicated, to correspond
with the grand fountain-heads of the rivers
they overshadowed. The Sierra Cathedral, to
the south of camp, was overshadowed like
Sinai. Never before noticed so fine a union of
rock and cloud in form and color and substance,
drawing earth and sky together as one; and
so human is it, every feature and tint of color
goes to one's heart, and we shout, exulting in
wild enthusiasm as if all the divine show were
our own. More and more, in a place like this,
we feel ourselves part of wild Nature, kin to
everything. Spent most of the day high up
on the north rim of the valley, commanding
views of the clouds in all their red glory spreading their wonderful light over all the basin,
while the rocks and trees and small Alpine

plants at my feet seemed hushed and thoughtful, as if they also were conscious spectators
of the glorious new cloud-world.

Here and there, as I plodded farther and
higher, I came to small garden-patches and
ferneries just where one would naturally de-
cide that no plant-creature could possibly live.
But, as in the region about the head of Mono
Pass and the top of Dana, it was in the wildest, highest places that the most beautiful
and tender and enthusiastic plant-people were
found. Again and again, as I lingered over
these charming plants, I said, How came you
here? How do you live through the winter?
Our roots, they explained, reach far down the
joints of the summer-warmed rocks, and beneath our fine snow mantle killing frosts cannot reach us, while we sleep away the dark
half of the year dreaming of spring.

Ever since I was allowed entrance into these
mountains I have been looking for cassiope,
said to be the most beautiful and best loved of
the heathworts, but, strange to say, I have not
yet found it. On my high mountain walks
I keep muttering, "Cassiope, cassiope." This
name, as Calvinists say, is driven in upon me,
notwithstanding the glorious host of plants
that come about me uncalled as soon as I show
myself. Cassiope seems the highest name of

all the small mountain-heath people, and as if conscious of her worth, keeps out of my way. I must find her soon, if at all this year.

September 4. All the vast sky dome is clear, filled only with mellow Indian summer light. The pine and hemlock and fir cones are nearly ripe and are falling fast from morning to night, cut off and gathered by the busy squirrels. Almost all the plants have matured their seeds, their summer work done; and the summer crop of birds and deer will soon be able to follow their parents to the foothills and plains at the approach of winter, when the snow begins to fly.

September 5. No clouds. Weather cool, calm, bright as if no great thing was yet ready to be done. Have been sketching the North Tuolumne Church. The sunset gloriously colored.

September 6. Still another perfectly cloudless day, purple evening and morning, all the middle hours one mass of pure serene sunshine. Soon after sunrise the air grew warm, and there was no wind. One naturally halted to see what Nature intended to do. There is a suggestion of real Indian summer in the hushed brooding, faintly hazy weather. The yellow atmosphere, though thin, is still plainly of the same general character as that of eastern

Indian summer. The peculiar mellowness is perhaps in part caused by myriads of ripe spores adrift in the sky.

Mr. Delaney now keeps up a solemn talk about the need of getting away from these high mountains, telling sad stories of flocks that perished in storms that broke suddenly into the midst of fine innocent weather like this we are now enjoying. "In no case," said he, "will I venture to stay so high and far back in the mountains as we now are later than the middle of this month, no matter how warm and sunny it may be." He would move the flock slowly at first, a few miles a day until the Yosemite Creek basin was reached and crossed, then while lingering in the heavy pine woods should the weather threaten he could hurry down to the foothills, where the snow never falls deep enough to smother a sheep. Of course I am anxious to see as much of the wilderness as possible in the few days left me, and I say again, — May the good time come when I can stay as long as I like with plenty of bread, far and free from trampling flocks, though I may well be thankful for this generous foodful inspiring summer. Anyhow we never know where we must go nor what guides we are to get, — men, storms, guardian angels, or sheep. Perhaps almost everybody in the least natural is guarded more than he is ever aware of. All the wilderness seems to be full of tricks and plans to drive and draw us up into God's Light.

Have been busy planning, and baking bread for at least one more good wild excursion among the high peaks, and surely none, however hopefully aiming at fortune or fame, ever felt so gloriously happily excited by the outlook.

September 7. Left camp at daybreak and made direct for Cathedral Peak, intending to strike eastward and southward from that point among the peaks and ridges at the heads of the Tuolumne, Merced, and San Joaquin Rivers. Down through the pine woods I made my way, across the Tuolumne River and meadows, and up the heavily timbered slope forming the south boundary of the upper Tuolumne basin, along the east side of Cathedral Peak, and up to its topmost spire, which I reached at noon, having loitered by the way to study the fine trees — two-leaved pine, mountain pine, albicaulis pine, silver fir, and the most charming, most graceful of all the evergreens, the mountain hemlock. High, cool, late-flowering meadows also detained me, and lakelets and avalanche tracks and huge quarries of moraine rocks above the forests.

All the way up from the Big Meadows to the base of the Cathedral the ground is covered with moraine material, the left lateral moraine of the great glacier that must have completely filled this upper Tuolumne basin. Higher there are several small terminal moraines of residual glaciers shoved forward at right angles against the grand simple lateral of the main Tuolumne Glacier. A fine place to study mountain sculpture and soil making. The view from the Cathedral Spires is very fine and telling in every direction. Innumerable peaks, ridges, domes, meadows, lakes, and woods; the forests extending in long curving lines and broad fields wherever the glaciers have left soil for them to grow on, while the sides of the highest mountains show a straggling dwarf growth clinging to rifts in the rocks apparently independent of soil. The dark heath-like growth on the Cathedral roof I found to be dwarf snow-pressed albicaulis pine, about three or four feet high, but very old looking. Many of them are bearing cones, and the noisy Clarke crow is eating the seeds, using his long bill like a woodpecker in digging them out of the cones. A good many flowers are still in bloom about the base of the peak, and even on the roof among the little pines, especially a woody yellow-flowered eri-

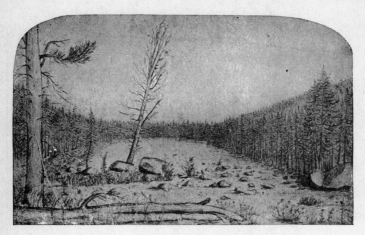

GLACIER MEADOW STREWN WITH MORAINE
BOULDERS, 10,000 FEET ABOVE THE SEA
(NEAR MOUNT DANA)

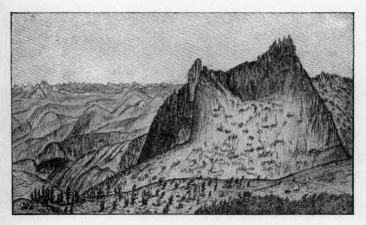

FRONT OF CATHEDRAL PEAK

ogonum and a handsome aster. The body of
the Cathedral is nearly square, and the roof
slopes are wonderfully regular and symmetrical, the ridge trending northeast and southwest. This direction has apparently been
determined by structure joints in the granite.
The gable on the northeast end is magnificent
in size and simplicity, and at its base there is
a big snow-bank protected by the shadow of
the building. The front is adorned with many
pinnacles and a tall spire of curious work-
manship. Here too the joints in the rock are
seen to have played an important part in determining their forms and size and general arrangement. The Cathedral is said to be about
eleven thousand feet above the sea, but the
height of the building itself above the level
of the ridge it stands on is about fifteen hundred feet. A mile or so to the westward there
is a handsome lake, and the glacier-polished
granite about it is shining so brightly it is not
easy in some places to trace the line between
the rock and water, both shining alike. Of
this lake with its silvery basin and bits of
meadow and groves I have a fine view from
the spires; also of Lake Tenaya, Cloud's Rest
and the South Dome of Yosemite, Mount Starr
King, Mount Hoffman, the Merced peaks,
and the vast multitude of snowy fountain

peaks extending far north and south along the axis of the range. No feature, however, of all the noble landscape as seen from here seems more wonderful than the Cathedral itself, a temple displaying Nature's best masonry and sermons in stones. How often I have gazed at it from the tops of hills and ridges, and through openings in the forests on my many short excursions, devoutly wondering, admiring, longing! This I may say is the first time I have been at church in California, led here at last, every door graciously opened for the poor lonely worshiper. In our best times everything turns into religion, all the world seems a church and the mountains altars. And lo, here at last in front of the Cathedral is blessed cassiope, ringing her thousands of sweet-toned bells, the sweetest church music I ever enjoyed. Listening, admiring, until late in the afternoon I compelled myself to hasten away eastward back of rough, sharp, spiry, splintery peaks, all of them granite like the Cathedral, sparkling with crystals — feldspar, quartz, hornblende, mica, tourmaline. Had a rather difficult walk and creep across an immense snow and ice cliff which gradually increased in steepness as I advanced until it was almost impassable. Slipped on a dangerous place, but managed to stop by digging my heels into the thawing surface just on the brink of a yawning ice gulf. Camped beside a little pool and a group of crinkled dwarf pines; and as I sit by the fire trying to write notes the shallow pool seems fathomless with the infinite starry heavens in it, while the onlooking rocks and trees, tiny shrubs and daisies and sedges, brought forward in the fire-glow, seem full of thought as if about to speak aloud and tell all their wild stories. A marvelously impressive meeting in which every one has something worth while to tell. And beyond the fire-beams out in the solemn darkness, how impressive is the music of a choir of rills singing their way down from the snow to the river! And when we call to mind that thousands of these rejoicing rills are assembled in each one of the main streams, we wonder the less that our Sierra rivers are songful all the way to the sea.

About sundown saw a flock of dun grayish sparrows going to roost in crevices of a crag above the big snow-field. Charming little mountaineers! Found a species of sedge in flower within eight or ten feet of a snow-bank. Judging by the looks of the ground, it can hardly have been out in the sunshine much longer than a week, and it is likely to be buried again in fresh snow in a month or so, thus making a winter about ten months long, while spring, summer, and autumn are crowded and hurried into two months. How delightful it is to be alone here! How wild everything is — wild as the sky and as pure! Never shall I forget this big, divine day — the Cathedral and its thousands of cassiope bells, and the landscapes around them, and this camp in the gray crags above the woods, with its stars and streams and snow.

September 8. Day of climbing, scrambling, sliding on the peaks around the highest source of the Tuolumne and Merced. Climbed three of the most commanding of the mountains, whose names I don't know; crossed streams and huge beds of ice and snow more than I could keep count of. Neither could I keep count of the lakes scattered on tablelands and in the cirques of the peaks, and in chains in the cañons, linked together by the streams — a tremendously wild gray wilderness of hacked, shattered crags, ridges, and peaks, a few clouds drifting over and through the midst of them as if looking for work. In general views all the immense round landscape seems raw and lifeless as a quarry, yet the most charming flowers were found rejoicing in countless nooks and garden-like patches everywhere. I must have done three or four days' climbing work in this

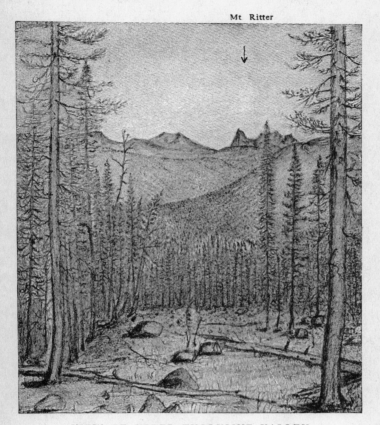

Mt Ritter

VIEW OF UPPER TUOLUMNE VALLEY

one. Limbs perfectly tireless until near sundown, when I descended into the main upper Tuolumne valley at the foot of Mount Lyell, the camp still eight or ten miles distant. Going up through the pine woods past the Soda Springs Dome in the dark, where there is much fallen timber, and when all the excitement of seeing things was wanting, I was tired. Arrived at the main camp at nine o'clock, and soon was sleeping sound as death.

CHAPTER XI

BACK TO THE LOWLANDS

September 9. Weariness rested away and I feel eager and ready for another excursion a month or two long in the same wonderful wilderness. Now, however, I must turn toward the lowlands, praying and hoping Heaven will shove me back again.

The most telling thing learned in these mountain excursions is the influence of cleavage joints on the features sculptured from the general mass of the range. Evidently the denudation has been enormous, while the inevitable outcome is subtle balanced beauty. Comprehended in general views, the features of the wildest landscape seem to be as harmoniously related as the features of a human face. Indeed, they look human and radiate spiritual beauty, divine thought, however covered and concealed by rock and snow.

Mr. Delaney has hardly had time to ask me how I enjoyed my trip, though he has facilitated and encouraged my plans all summer, and declares I'll be famous some day, a kind guess that seems strange and incredible to a wandering wilderness-lover with never a thought or dream of fame while humbly trying to trace and learn and enjoy Nature's lessons.

The camp stuff is now packed on the horses, and the flock is headed for the home ranch. Away we go, down through the pines, leaving the lovely lawn where we have camped so long. I wonder if I'll ever see it again. The sod is so tough and close it is scarcely at all injured by the sheep. Fortunately they are not fond of silky glacier meadow grass. The day is perfectly clear, not a cloud or the faintest hint of a cloud is visible, and there is no wind. I wonder if in all the world, at a height of nine thousand feet, weather so steadily, faithfully calm and bright and hospitable may anywhere else be found. We are going away fearing destructive storms, though it is difficult to conceive weather changes so great.

Though the water is now low in the river, the usual difficulty occurred in getting the flock across it. Every sheep seemed to be invincibly determined to die any sort of dry death rather than wet its feet. Carlo has learned the sheep business as perfectly as the best shepherd, and it is interesting to watch his intelligent efforts to push or frighten the silly creatures into the water. They had to be fairly crowded and shoved over the bank; and when at last one crossed because it could not push its way back, the whole flock suddenly plunged in headlong together, as if the river was the only desirable part of the world. Aside from mere money profit one would rather herd wolves than sheep. As soon as they clambered up the opposite bank, they began baaing and feeding as if nothing unusual had happened. We crossed the meadows and drove slowly up the south rim of the valley through the same woods I had passed on my way to Cathedral Peak, and camped for the night by the side of a small pond on top of the big lateral moraine.

September 10. In the morning at daybreak not one of the two thousand sheep was in sight. Examining the tracks, we discovered that they had been scattered, perhaps by a bear. In a few hours all were found and gathered into one flock again. Had fine view of a deer. How graceful and perfect in every way it seemed as compared with the silly, dusty, tousled sheep! From the high ground hereabouts had another grand view to the northward — a heaving, swelling sea of domes and round-backed ridges fringed with pines, and bounded by innumerable sharp-pointed peaks, gray and barren-looking, though so full of beautiful life. Another day of the calm, cloudless kind, purple in the morning and evening. The evening glow has been very marked for the last two or three weeks. Perhaps the "zodiacal light."

September 11. Cloudless. Slight frost. Calm. Fairly started downhill, and now are camped at the west end meadows of Lake Tenaya — a charming place. Lake smooth as glass, mirroring its miles of glacier-polished pavements and bold mountain walls. Find aster still in flower. Here is about the upper limit of the dwarf form of the goldcup oak, — eight thousand feet above sea-level, — reaching about two thousand feet higher than the California black oak (*Quercus Californica*). Lovely evening, the lake reflections after dark marvelously impressive.

September 12. Cloudless day, all pure sungold. Among the magnificent silver firs once more, within two miles of the brink of Yosemite, at the famous Portuguese bear camp. Chaparral of goldcup oak, manzanita, and ceanothus abundant hereabouts, wanting about the Tuolumne meadows, although the elevation is but little higher there. The two-leaved pine, though far more abundant about the Tuolumne meadow region, reaches its greatest size on stream-sides hereabouts and around meadows that are rather boggy. All the best dry ground is taken by the magnificent silver fir, which here reaches its greatest size and forms a well-defined belt. A glorious tree. Have fine bed of its boughs to-night.

September 13. Camp this evening at Yosemite Creek, close to the stream, on a little sand flat near our old camp-ground. The vegetation is already brown and yellow and dry; the creek almost dry also. The slender form of the two-leaved pine on its banks is, I think, the handsomest I have anywhere seen. It might easily pass at first sight for a distinct species, though surely only a variety (*Murrayana*), due to crowded and rapid growth on good soil. The yellow pine is as variable, or perhaps more so. The form here and a thousand feet higher, on crumbling rocks, is broad branching, with closely furrowed, reddish bark, large cones, and long leaves. It is one of the hardiest of pines, and has wonderful vitality. The tassels of long, stout needles shining silvery in the sun, when the wind is blowing them all in the same direction, is one of the most splendid spectacles these glorious Sierra forests have to show. This variety of *Pinus ponderosa* is regarded as a distinct species, *Pinus Jeffreyi*, by some botanists. The basin of this famous Yosemite stream is extremely rocky, — seems fairly to be paved with domes like a street with big cobblestones. I wonder if I shall ever be allowed to explore it. It draws me so strongly, I would make any sacrifice to try to read its lessons. I thank God for this glimpse of it. The charms of these mountains are beyond all common reason, unexplainable and mysterious as life itself.

September 14. Nearly all day in magnificent fir forest, the top branches laden with superb erect gray cones shining with beads of pure balsam. The squirrels are cutting them off at a great rate. Bump, bump, I hear them falling, soon to be gathered and stored for winter bread. Those that chance to be left by the industrious harvesters drop the scales and bracts when fully ripe, and it is fine to see the purple-winged seeds flying in swirling, merry-looking flocks seeking their fortunes. The bole and dead limbs of nearly every tree in the main forest-belt are ornamented by conspicuous tufts and strips of a yellow lichen.

Camped for the night at Cascade Creek, near the Mono Trail crossing. Manzanita berries now ripe. Cloudiness to-day about .10. The sunset very rich, flaming purple and crimson showing gloriously through the aisles of the woods.

September 15. The weather pure gold, cloudiness about .05, white cirrus flects and pencilings around the horizon. Move two or three miles and camp at Tamarack Flat. Wandering in the woods here back of the pines which

bound the meadows, I found very noble specimens of the magnificent silver fir, the tallest
about two hundred and forty feet high and
five feet in diameter four feet from the ground.

September 16. Crawled slowly four or five
miles to-day through the glorious forest to
Crane Flat, where we are camped for the night.
The forests we so admired in summer seem
still more beautiful and sublime in this mellow
autumn light. Lovely starry night, the tall,
spiring tree-tops relieved in jet black against
the sky. I linger by the fire, loath to go to bed.

September 17. Left camp early. Ran over
the Tuolumne divide and down a few miles to
a grove of sequoias that I had heard of, di-
rected by the Don. They occupy an area of
perhaps less than a hundred acres. Some of the
trees are noble, colossal old giants, surrounded by
magnificent sugar pines and Douglas spruces.
The perfect specimens not burned or broken are
singularly regular and symmetrical, though not
at all conventional, showing infinite variety in
general unity and harmony; the noble shafts
with rich purplish brown fluted bark, free of
limbs for one hundred and fifty feet or so, ornamented here and there with leafy rosettes;
main branches of the oldest trees very large,
crooked and rugged, zigzagging stiffly outward
seemingly lawless, yet unexpectedly stooping

just at the right distance from the trunk and dissolving in dense bossy masses of branchlets, thus making a regular though greatly varied outline, — a cylinder of leafy, outbulging spray masses, terminating in a noble dome, that may be recognized while yet far off upheaved against the sky above the dark bed of pines and firs and spruces, the king of all conifers, not only in size but in sublime majesty of behavior and port. I found a black, charred stump about thirty feet in diameter and eighty or ninety feet high — a venerable, impressive old monument of a tree that in its prime may have been the monarch of the grove; seedlings and saplings growing up here and there, thrifty and hopeful, giving no hint of the dying out of the species. Not any unfavorable change of climate, but only fire, threatens the existence of these noblest of God's trees. Sorry I was not able to get a count of the old monument's annual rings.

Camp this evening at Hazel Green, on the broad back of the dividing ridge near our old camp-ground when we were on the way up the mountains in the spring. This ridge has the finest sugar-pine groves and finest manzanita and ceanothus thickets I have yet found on all this wonderful summer journey.

September 18. Made a long descent on the

south side of the divide to Brown's Flat, the grand forests now left above us, though the sugar pine still flourishes fairly well, and with the yellow pine, libocedrus, and Douglas spruce, makes forests that would be considered most wonderful in any other part of the world.

The Indians here, with great concern, pointed to an old garden patch on the flat and told us to keep away from it. Perhaps some of their tribe are buried here.

September 19. Camped this evening at Smith's Mill, on the first broad mountain bench or plateau reached in ascending the range, where pines grow large enough for good lumber. Here wheat, apples, peaches, and grapes grow, and we were treated to wine and apples. The wine I did n't like, but Mr. Delaney and the Indian driver and the shepherd seemed to think the stuff divine. Compared to sparkling Sierra water fresh from the heavens, it seemed a dull, muddy, stupid drink. But the apples, best of fruits, how delicious they were — fit for gods or men.

On the way down from Brown's Flat we stopped at Bower Cave, and I spent an hour in it — one of the most novel and interesting of all Nature's underground mansions. Plenty of sunlight pours into it through the leaves of the four maple trees growing in its mouth, illuminating its clear, calm pool and marble chambers, — a charming place, ravishingly beautiful, but the accessible parts of the walls sadly disfigured with names of vandals.

September 20. The weather still golden and calm, but hot. We are now in the foot-hills, and all the conifers are left behind except the gray Sabine pine. Camped at the Dutch Boy's Ranch, where there are extensive barley fields now showing nothing save dusty stubble.

September 21. A terribly hot, dusty, sunburned day, and as nothing was to be gained by loitering where the flock could find nothing to eat save thorny twigs and chaparral, we made a long drive, and before sundown reached the home ranch on the yellow San Joaquin plain.

September 22. The sheep were let out of the corral one by one, this morning, and counted, and strange to say, after all their adventurous wanderings in bewildering rocks and brush and streams, scattered by bears, poisoned by azalea, kalmia, alkali, all are accounted for. Of the two thousand and fifty that left the corral in the spring lean and weak, two thousand and twenty-five have returned fat and strong. The losses are: ten killed by bears, one by a rattlesnake, one that had to be killed

MY FIRST SUMMER IN THE SIERRA

after it had broken its leg on a boülder slope, and one that ran away in blind terror on being accidentally separated from the flock, — thirteen all told. Of the other twelve doomed never to return, three were sold to ranchmen and nine were made camp mutton.

Here ends my forever memorable first High Sierra excursion. I have crossed the Range of Light, surely the brightest and best of all the Lord has built; and rejoicing in its glory, I gladly, gratefully, hopefully pray I may see it again.

THE END

INDEX

INDEX

Butler, Prof. J. D., strange experience of Muir with, 178–91.

Butterflies, 160.

Calochortus albus, 17.

Camping, in the foothills, 10, 11; on the North Fork of the Merced, 32–74; at Tamarack Flat, 99; in the Yosemite, 122; near Soda Springs, 201, 229; alone, in Bloody Cañon, 220–22; on the Tuolumne, 232–53.

Cañon Creek, 223.

Carlo, St. Bernard dog, with Muir in the Sierra, 5, 6, 43, 57, 59, 60, 62, 123, 124, 154, 181, 192, 193; afraid of bears, 116, 135; runs away, 232, 233, 255.

Cascade Creek, 104, 259.

Cassiope, 244, 250.

Cathedral Peak, 154, 212, 231, 247, 250; well named, 146; a majestic temple, 198; view from, 248; height, 249.

Cedar, incense (*Libocedrus decurrens*), 20, 21, 93.

Chamæbatia foliolosa, 33, 34.

Chinaman, shepherd's helper, 6, 9.

Chipmunk, in the Sierra, 171, 172.

Cleavage joints, 254.

Clouds, 56, 73, 147, 148, 242, 243; sky mountains, 18, 19, 37, 39, 61, 133, 144, 145.

Coffee, 82.

Corylus rostrata, 65.

Coulterville, 9, 17, 19.

Crane Flat, 90, 92, 93, 260.

Crows, 9, 248.

Crystals, radiant, 153, 250; frost, 234, 236.

Daisy, blue arctic, 218.

Deer, black-tailed, 142.

Delaney, Mr., sheep-owner, 6, 12, 25, 27, 36, 83, 103, 104, 112–14, 194, 206, 233, 238, 246, 254, 262; engages Muir to go with his flock to the Sierra, 4, 5; describes David Brown's method of bear-hunting, 28–30; talks of bears in general, 107, 108; a big-hearted Irishman, 214.

Dendromecon rigidum, 39.

Devil's slides, 150.

Dogwood, Nuttall's flowering, 64.

Dome Creek, 121.

Don Quixote, nickname for Mr. Delaney, 6, 12.

Elymus (wild rye), 226.

Emerald Pool, 189.

Eskimo, 69.

Fawn, baby, 232.

Ferns, 40, 41.

Fir, silver, 90–93, 98, 105, 257; cones, 91, 167, 168, 259; size, 143, 161, 162, 166, 260; age, 166, 167; leaves, 167.

Fire, in woods, 19, 202, 203.

Fishes, none in high Sierra lakes, 200.

Flicker, 173.

Floods, 48.

266

INDEX

Flowers, in Merced Valley, 33, 35, 36, 40, 58; at Crane Flat, 92, 93, 94; on Yosemite Creek, 109, 110; on Hoffman Range, 151, 152, 158, 160, 196; in Tuolumne Meadows, 199, 203; in Bloody Cañon, 218, 224, 225, 228, 230.

Flowing, everything is, 236.

Food, of bears, 28, 29, 46, 192; of squirrels, 18, 69, 74, 168; of Indians, 12, 46, 70, 226–28.

Foothills, 3–31.

Frogs, in the highest lakes, 200.

Frost, crystals, 234, 236.

Gallflies, 170.

Glacial action, 101, 102, 196, 197, 200, 202, 203, 205, 208, 215, 216, 224, 240, 248.

Glacier meadows, 229, 230.

Gold region, 55, 56; mines near Mono Lake, 105.

Grasshopper, a queer fellow, 139–41.

Greeley's Mill, 17, 20.

Grouse, blue or dusky, 175, 176.

Half-Dome, or South Dome, 117, 122, 129.

Hare, 9.

Hare, little chief, 154, 155.

Hazel, beaked, 65.

Hazel Creek, 89.

Hazel Green, 87, 261.

Heat, in the foothills, 8.

Hemlock, mountain (*Tsuga Mertensiana*), 151, 247.

Hogs, 108.

Horseshoe Bend, 13, 19.

House-fly, on North Dome, 138, 139; on Mount Hoffman, 169.

Hutchings, Mrs., landlady, 182.

Illilouette, 189.

Indian Basin, 121.

Indian Cañon, 115, 122, 181, 186, 187.

Indian Creek, 208.

Indians, Digger, 12, 30, 31, 262; shepherd's helper with Muir, 6, 9, 10, 86, 90; anteaters, 46; their power of escaping observation, 53, 54, 58; an old woman, 58, 59; Chief Tenaya, 165; a hunter, 205, 206; food, 206, 226, 227; a dirty band, 218, 219; women gathering wild rye, 226.

Ivy, poison, 26.

Jack, the shepherd's little dog, 62, 63.

Joe, Portuguese shepherd, 209, 210.

Juniper, Sierra (*Juniperus occidentalis*), 110, 163–65.

Lake Hoffman, 154.

Lake Tenaya, 153, 155, 165, 195–97, 257; Indian name, 166.

Landscape, sculpture of, 14; a glorious, 115, 116; features harmonious, 240, 254.

Liberty Cap, 183.

Libocedrus decurrens. See Cedar, incense.

267

INDEX

Lichens, 259.
Lightning, 15, 124, 125.
Lilies, 36, 37, 59, 60, 225.
Lilium pardalinum, 37.
Lilium parvum, 94, 95, 121.
Lily, twining, 50; on poison ivy, 26.
Lily, Washington, 103.
Linosyris, 20.
Lizards, 8, 41–43, 65.

Magpies, 9.
Mammoth Mountain,216,242.
Manzanita (*Arctostaphylos*), 88, 89; berries, 259.
Meadows, three kinds of, 158, 159; glacier, 229, 230.
Merced River, 189; North Fork of,25; camp on, 32–74.
Merced Valley, 13, 115.
Mono Desert, 226.
Mono Lake, 214, 226, 239; flowers around, 228.
Mono Trail, 104, 109, 115, 195–213.
Moon, startling effect of, 221, 222.
Moraine Lake, 224, 225.
Moraines, 102, 216, 224, 240, 248.
Mosquitoes, Sierra, 169.
Mount Dana, 199, 230, 233, 234, 239, 242.
Mount Gibbs, 199, 242.
Mount Hoffman, 230; height of, 149; watershed, 150; flowers, 151, 152, 158, 160; hemlocks and pines, 151, 152; crystals, 153; strange dove-colored bird, 176.
Mount Lyell, 198, 253.
Mutton, exclusive diet of, 76.

Neotoma, 71–73.
Nevada Cañon, 182.
Nevada Fall, 187, 188, 207.
North Dome,131,134; strange experience on, 178, 179.

Oak, blue (*Quercus Douglasii*), 8, 15.
Oak,California black (*Quercus Californica*), 15, 257.
Oak, dwarf (*Quercus chrysolepis*), 161.
Oak, goldcup, 50, 187, 257.
Oak, mountain live, 38.
Oak, poison, 26.
Oreortyx ricta, 174, 175.

Pictures, inadequate, 131.
Pika, 154, 155.
Pilot Peak Ridge, 32, 57, 65, 67, 84.
Pine, dwarf (*Pinus albicaulis*), 152, 248; as fuel, 221.
Pine, mountain (*Pinus monticola*), 152.
Pine, Sabine, 12, 13, 263; cones, 12.
Pine, silver, 52.
Pine, sugar, 17, 18, 51, 88, 90, 93; cones, 50.
Pine, two-leaved or tamarack, 99, 110, 162, 163, 257, 258.
Pine, yellow, 15, 51, 52, 88, 93, 258; cones, 17, 18.
Pino Blanco, 13.
Poppy, bush (*Dendromecon rigidum*), 39.
Porcupine Creek, 121, 206.
Portuguese shepherds, 206, 207, 208–10.
Pseudotsuga Douglasii, 93.
Pteris aquilina, 40, 41.

INDEX

Quail, mountain (*Oreortyx ricta*), 174, 175.
Quails, 9.
Quercus Califórnica, 15, 257.
Quercus chrysolepis, 161.
Quercus Douglasii, 8, 15.

Rabbits, cottontail, 9, 227.
Raindrop, history of, 125–27.
Range of Light, 236, 264.
Rat, wood (*Neotoma*), 71–73.
Rattlesnakes, 9; dog bitten by one, 63.
Rhus diversiloba. See Ivy, poison.
Robin, 173, 174, 218.
Rye, wild, 226.

Sandy, David Brown's dog, 27, 28, 30.
Saxifrage, giant (*Saxifraga peltata*), 35.
Sedge, 34, 35.
Seeds, 68.
Sequoia gigantea, 93; grove of, 260, 261.
Shadows, of leaves, 59; substantial looking, 233.
Sheep, Mr. Delaney's flock, 5, 8, 9, 11, 61, 64, 86, 87, 256, 263, 264; rate of travel, 7; camping, 10; poisoned by azalea, 22; profitable, 22; hoofed locusts, 56, 86; stray, 57; destructiveness of, 97, 195; crossing a creek, 111–14; have poor brain stuff, 114; raided by bears, 191, 192, 194; afraid of getting wet, 201, 202, 255.

Shepherd, degrading life of the Californian, 23; in Scotland, 24; the oriental, 24; bed and food, 80, 81.
Slate, metamorphic, 6, 8, 14, 34.
Smith's Mill, 262.
Soda Springs, 201, 229, 253.
South Dome, 122, 129.
Sparrows, 251.
Spiders, 53.
Spruce, Douglas, 93.
Squirrel, California gray, 69, 70.
Squirrel, Douglas, 18, 68–70, 96, 168.
Stropholirion Californicum. See Lily, twining.
Sunrise, in the Yosemite, 124.
Sunset, 53.

Tamarack Creek, 100, 102, 106.
Tamarack Flat, 90, 259.
Tea, 80, 82.
Telepathy, strange case of, 178–91.
Tenaya, Yosemite chief, 165.
Tenaya Creek, 156.
Three Brothers, 207.
Thunder, in the mountains, 122, 123, 125.
Tissiack. *See* Half-Dome.
Tourists, 98, 104, 190.
Trees and storm, 144.
Tuolumne Camp, 232–53.
Tuolumne Meadows, 198, 199.

Vaccinium, dwarf, 218.
Veratrum Californicum, 93, 94.

INDEX